Non-Equilibrium Entropy and Irreversibility

MATHEMATICAL PHYSICS STUDIES

A SUPPLEMENTARY SERIES TO
LETTERS IN MATHEMATICAL PHYSICS

VOLUME 5

Non-Equilibrium Entropy and Irreversibility

by

GÖRAN LINDBLAD

Department of Theoretical Physics,
Royal Institute of Technology, Stockholm, Sweden

D. Reidel Publishing Company

A MEMBER OF THE KLUWER ACADEMIC PUBLISHERS GROUP

Dordrecht / Boston / Lancaster

Library of Congress Cataloging in Publication Data

Lindblad, Göran, 1940–
    Non-equilibrium entropy and irreversibility.

    (Mathematical physics studies ; v. 5)
    Bibliography: p.
    Includes indexes.
    1. Entropy. 2. Irreversible processes. I. Title. II. Series.
QC318.E57L56   1983   536'.73   83-15953
ISBN 90-277-1640-4

Published by D. Reidel Publishing Company
P.O. Box 17, 3300 AA Dordrecht, Holland

Sold and distributed in the U.S.A. and Canada
by Kluwer Academic Publishers,
190 Old Derby Street, Hingham, MA 02043, U.S.A.

In all other countries, sold and distributed
by Kluwer Academic Publishers Group,
P.O. Box 322, 3300 AH Dordrecht, Holland

Printed in The Netherlands

TABLE OF CONTENTS

PREFACE

The problem of deriving irreversible thermodynamics from the reversible microscopic dynamics has been on the agenda of theoretical physics for a century and has produced more papers than can
be digested by any single scientist. Why add to this too long list
with yet another work? The goal is definitely not to give a general review of previous work in this field. My ambition is rather
to present an approach differing in some key aspects from the standard treatments, and to develop it as far as possible using rather
simple mathematical tools (mainly inequalities of various kinds).
However, in the course of this work I have used a large number of
results and ideas from the existing literature, and the reference
list contains contributions from many different lines of research.
As a consequence the reader may find the arguments a bit difficult
to follow without some previous exposure to this set of problems.

The idea which started this work is that the relation between
dynamics and thermodynamics can be based on the concepts of energy
and work, which are common to both fields. The work done by a
quantum or classical Hamiltonian system is defined through the response of the system to cyclic changes in a set of time-dependent
macroscopic external fields. The entropy of a state is then defined in such a way that it is simply related to the available
work. This means that it is described completely in terms of dynamical quantities. A consequence of this definition is that there
can be a unique, intrinsic entropy only if its value is constant
in time for a closed system. Hence, there is no intrinsic irreversibility in this formalism. Instead, there is a family of non-
trivial entropy functions, one for each set of thermodynamic pro-
esses allowed by the experimenter's control of the system through

the external fields.

The present approach is closely modelled on a type of experiment where the relaxation properties are probed by modern techniques which provide a high time resolution. The fact that such experiments were not possible at the time when non-equilibrium statistical mechanics was in its infancy has influenced its development up to the present. A result is that the standard types of rigorous mathematical treatment, as for example that given by ergodic theory, provide a rather tenous relation between the basic concepts and the experimentally accessible quantities. In this work I have tried to improve on this situation. There is, for instance no a priori information-theoretic interpretation of the entropy for non-equilibrium states, as this is not relevant for the given set of experiments.

Another basic idea behind the present scheme is that the second law of thermodynamics, if it is to have a universal validity, must be a tautology in a certain sense. Here the second law is based on a universal property, namely a form of causality which is introduced as a Markov property of the dynamical description. With such a wide interpretation of the second law it is not always true that the entropy will increase in time to the equilibrium value. It can even be argued that only the calculation of a finite rate of relaxation to equilibrium in a particular model is physically relevant, but this type of problem is beyond the present abstract formalism. All the same, the entropy functions given here define measures of the deviation from equilibrium and give a well-defined sense to the notion of relaxation to equilibrium for any given set of thermodynamic processes.

The detailed working out of this general scheme meets with a number of conceptual and mathematical difficulties, most of which are familiar from the standard approaches. There is, for instance, the recurrence paradox associated with the almost periodic properties of closed finite quantum systems. All of these problems are not solved in the detail which they merit. This means that mathematical rigour is lacking in several places, heuristic and qualitative arguments being used instead. However, I believe that the present approach improves the understanding of the approximations involved in a thermodynamic description of finite quantum systems. An important aspect here is the specification of the set of experiments to which the thermodynamic concepts are intended to apply, a set which must necessarily differ from that relevant for a microcopic description of the system. I find it likely that the program of giving a more rigorous foundation of this field can only be advanced by providing more realistic models of the experiments as well as a better qualitative understanding of the Hamiltonian dynamics of large systems.

Acknowledgements

The research behind this work was supported by the Swedish Natural Science Research Council. I wish to thank Professor W. Thirring and Dr. O. Berg for comments and some references, Dr. C. Obcemea and Dr. W.A. Majewski for the correction of a number of errors in a preliminary version.

CHAPTER 1. INTRODUCTION AND SUMMARY.

Two related conceptual problems in the foundations of statistical mechanics concern the derivation of the irreversibility of the observed macroscopic behaviour from the reversible microscopic laws of motion and the definition of an entropy function on non-equilibrium states. The connection between these two concepts lies in the desired property of the entropy function to be non-decreasing under all physically allowed evolutions of a closed system, including the response of the system to changes in boundary conditions. Furthermore it should increase to the value corresponding to the Gibbs canonical state of the appropriate energy if the system is left to itself.

The number of investigations in this field since the pioneer work of Boltzmann is so immense that no review can be attempted here. For background material the reader is referred to the recent and comprehensive review by Penrose [1] and the book by Davies[2].

One motivation behind a recent increase of interest in this set of problems may be the following facts. Advances in experimental techniques have made it possible to study relaxation phenomena on a very short time scale in systems which are subjected to strong and rapidly varying external fields. Examples include resonance and relaxation phenomena in nuclear magnetism and quantum optics. There has been numerous attempts to apply thermodynamic concepts in the interpretation of these experiments. A case in point is the introduction of the notion of *spin temperature* to describe energy exchange in spin systems [3,4].

Some experiments, like those displaying *spin echoes* seem to indicate that the concepts of temperature and relaxation to equilibrium may have a restricted validity for these systems. They show that part of the "relaxation" can be reversed if the system is subjected to a suitably chosen time-dependent magnetic field [5]. This is true despite the fact that the spin system is a genuine interacting many-body system, but it requires that it can be isolated from the crystal lattice during the experiment. In fact the observed spin-spin relaxation time of the magnetic polarization in a sample of CaF_2 at sufficiently low temperature can in a similar way be prolonged by a factor ($> 10^3$) which seems to be limited only by the performance of the experimental apparatus [6].

This possibility of reversing or suppressing a seemingly irreversible evolution has been compared with the velocity reversal paradox familiar from discussions of Boltzmann's H-theorem (even a *Loschmidt demon* performing this feat was invoked [5]). The spin echo experiments, however, do not involve antiunitary transformations like velocity reversal or observations of the microscopic state of the system. The operations performed on the system are simply unitary transformations generated by strong magnetic fields via the Zeeman Hamiltonian, and they are (at least in the standard picture) reversible and entropy-preserving. The possibility of restoring the initial state through the action of macroscopic fields forces me to the conclusion that there is a lack of intrinsic irreversibility in this type of system, at least to the limit of precision of the experiments.

It seems to me that the type of experiments referred to above rather than contradicting thermodynamics in any sense do in fact fit rather nicely into a scheme of thermodynamics of irreversible processes which starts from the reversible microscopic dynamics. The dynamic transformations generated by the time-dependent fields in the spin-echo experiments turn out to be the natural generalization of the Carnot cycles of equilibrium thermodynamics to the

non-equilibrium situation. As this scheme departs in some basic
aspects from the standard approach used in discussions of the
transition from microscopic dynamics to irreversible macroscopic
phenomena, it is perhaps justified to outline the philosophy behind it at some length.

Classical thermodynamics deals with a restricted set of phenomena, which can be described in terms of concepts like energy,
work and heat. The interaction of thermodynamic systems is reduced to energy exchange (if particle number is conserved). The
energy of a system is changed as work is performed on it by variation of external constraints and fields. In this way a certain
class of thermodynamic processes is defined. The entropy is a
state function which tells us which equilibrium states can be
reached from a given state through these processes. The isentropic
(reversible) processes have the important property of having optimal efficiency in producing work out of the system through
cyclic changes in the external fields.

Thermodynamics may thus be described as the theory of energy
changes in systems with time-dependent external fields. From this
point of view it is a bit surprising that most of the work that
has been done on the derivation of irreversibility and the justification of thermodynamics from microscopic dynamics has been restricted to systems with a prescribed time-homogenous evolution.
In adopting this restriction several aspects are lost. First, the
possibility of considering the reversal of the evolution through
a suitable time-dependent external field is neglected. Consequently the concept of irreversibility must lose much of its
proper meaning. Secondly, the concepts of work and heat can not
enter the theory. Hence there is some difficulty in identifying
the observables of the microscopic theory with thermodynamic quantities. Often the thermodynamic interpretation is based on an
"entropy" which is introduced ad hoc with the goal of showing its

monotonic behaviour under the evolution of the state, but which
has a priori no relation with the available work. Finally, the
equilibrium states reached through the irreversible evolution can
not be characterized in the way which is most natural in thermodynamics. This is through Kelvin's form of the second law which
states that it is impossible to perform a cyclic process on a system in equilibrium in which the only effect is the extraction of
work.

The present work starts from the idea that the connection
between microscopic dynamics and thermodynamics should be based
on the concepts which are common to both, namely *energy* and *work*.
The model is as follows: A quantum or classical system is given
where the Hamiltonian is a function of certain macroscopic external fields which the experimenter can vary at will. For any evolution defined by a time-dependent Hamiltonian in this family the
work performed by the system on the fields is well-defined.

In order to allow reproducible experiments there must exist
the means to prepare the system in one of a number of well-defined
initial states. For the purpose of thermodynamics the natural
thing is to prepare the system in an equilibrium (Gibbs) state.
There is one such for each pair of values of Hamiltonian and temperature (Chapter 3). However, it is not true that the state of a
general system will approach equilibrium if the system is left to
itself. Thus it is necessary to introduce a set of idealized heat
baths which can perform this preparation. They are represented by
infinite quasifree systems with good relaxation properties (Chapter 4). If the system is initially in an equilibrium state, an infinitely slow change of the external fields, while the system is in
contact with a succession of different.heat baths, will give a reversible process. The state of the system is then assumed always
to be the equilibrium state defined by the instantaneous values of
the fields and the temperature of the reservoir (Chapter 5).

4

Again starting from one of the equilibrium states, a set of reproducible non-equilibrium states is generated by the semigroup of evolutions defined by the available time-dependent Hamiltonians. With a cyclic change in the Hamiltonian, work is performed on the system in order to create the non-equilibrium states.

Given a reproducible non-equilibrium state, we can ask for the maximum work that can be recovered from the system when the Hamiltonian is again varied in a cyclic fashion with a prescribed origin. If the state is an equilibrium state (in general corresponding to a Hamiltonian not equal to the origin of the cycles), then one optimal work cycle is composed of reversible processes. In general optimal work cycles are not reversible, but they are still isentropic processes in this formalism. The entropy of a non-equilibrium state is defined as the infimum of the entropies of the equilibrium states which can be reached from the given state using the semigroup of evolutions generated by the Hamiltonian dynamics and the interaction with the heat baths. The entropy so defined is directly related to the available work and it has other desirable properties like being non-decreasing in time. It is in general not trivially constant in time, but convergence to the equilibrium value can not be proved in the present abstract formalism.

The entropy function is not unique. Instead there is a family of such functions, one for each set of thermodynamic processes allowed by the experimenter's control of the dynamics of the system through the external fields. This scheme is in line with the philosophy described by Jaynes' dictum: "Entropy is a property, not of the physical system, but of the particular experiments you or I choose to perform on it" [7]. Jaynes is the chief advocate of the maximum entropy formalism, where the value of the entropy is determined by the information obtained about the state of the system. Here, however, the difference between two entropy functions is not due to the degree of the experimenter's knowledge of the state, but rather to the different specifications of the set of experiments which he can perform on the system in the future. The entropy functions obtained here are not of the maximum entropy variety as a comparision of the two concepts will show in Chapter 12.

In order to explain clearly the reasons for the final choice of definition of the entropy functions and to show the difficulties which turn up on the way, I have chosen to give a sequence of different treatments applicable to different types of systems. An outline will now be given.

For finite closed systems (treated in Chapter 6) it is not possible to give an absolute significance to the concept of irreversibility, or to specify exactly what properties such a system should have in order to show irreversibility in a given approximation. The problems are due to the recurrences which are inevitable in finite bounded quantum systems. The formalism adopted here, which starts from the concept of work cycles, is particularly suited for a discussion of this aspect. This is due to the fact that the time homogeneity is broken by the choice of a starting time and a finite duration for the work cycles. It is then possible to define what should be meant by recurrences in this context and how they influence the value of the entropy. As little is known about the recurrences for large systems, this type of argument must unfortunately remain largely qualitative.

The finite duration of work cycles is not only necessary from such esoteric considerations as the finite lifetime of the universe. A much more stringent restriction on the experiments is given by the impossibility of isolating a system in a state far from equilibrium from the external world for any length of time. This point is discussed in Chapter 8. There it is argued that

quantum systems can be highly sensitive to perturbations which
are much too weak to cause a significant transfer of energy. The
argument is by analogy with classical dynamical systems where a
similar sensitivity seems to be characteristic of systems with
exponentially diverging trajectories (Appendix B). It is suggested that such properties, effective during a finite time interval, can be compatible with the recurrence properties of finite
quantum systems. The problem of justifying this conjecture is related to that of defining a notion of stochastic dynamics for
quantum systems. Here much remains to be done in order to obtain
a satisfactory understanding.

When the system can exchange energy with the exterior world
(apart from the given macroscopic fields), say with a heat bath
of a given temperature, then the formalism of open systems is
applicable (Chapter 7). The reduced dynamics of the small open
system, which describes the relaxation due to the action of the
heat bath, is then irreversible in a well-defined sense. The
simplest type of reduced dynamics of this type is given by the
Markovian master equations (Appendix A). It would be easy to believe that the thermodynamics of such models should be straightforward and perfect. In fact, the opposite is true for quantum
systems. The approximations which make it possible to neglect the
correlations between the open system and the heat bath also de-
stroys the thermodynamic consistency: An infinite amount of work
can be extracted from the system plus heat bath if the work cycles
can be chosen arbitrarily (Appendix A). This just means that the
derivation of the master equation must break down for this type
of time-dependent forces. In fact there seems to be no simple type
of memoryless reduced quantum dynamics such that the second law
applies to the system plus heat bath (except for infinite temperature heat baths). On the other hand, it is essential for the definition of the entropy and other thermodynamic state functions that
the dynamic description is Markovian in the sense that the state

of the system at any instant defines the future evolution of the state.

It seems that the only dynamics which is completely consistent with thermodynamics and useful for defining the thermodynamic state functions is the Hamiltonian type. The action of the heat baths on the system has to be restricted to the gentle time-dependence of reversible processes where the contradictions mentioned above will not be evident. For the same reason the uncontrolled perturbations due to the rest of the external world can be modelled by a Markovian process only ᴜₛ long as there is no significant energy exchange between the system and the reservoir. This is the scheme adopted in Chapter 10.

The entropy functions defined in Chapter 10 are in general different from the information (statistical) entropy. The information entropy is a lower bound for the entropy functions, and on equilibrium states they all coincide. This lack of identity is due to the fact that finding the optimal work cycles and the available work involves solving the equations of motion. This contrasts with the reversible Carnot cycles which are the optimal cycles for equilibrium states and which are universal and independent of the dynamics. For this reason the entropy functions defined here have no real predictive value. Their usefulness is on the conceptual level, as they are a tool for the understanding of the nature of irreversibility. In contrast the information entropy of equilibrium states can in principle be calculated from the Hamiltonian without solving the dynamics. Thus it has some predictive value. However, being constant under the Hamiltonian dynamics, it can not describe the irreversibility which is due to the complexity of a large closed system.

The relation between information and entropy is considered in more detail in Chapter 11. From the standpoint of the present

formalism this relation is highly indirect. This becomes evident
when the possibility of making measurements on the fluctuations
of the system is taken into account. The definition of the entropy
functions is such that they are not additive for interacting systems, and this is true even for states without statistical correlations between the systems. Thus the interaction of an observed
system and the measuring apparatus makes it difficult to assign
a unique value of the entropy to each of them. It is possible to
introduce a new entropy function for the system which takes into
account the set of possible measurements, or one for the system
plus the apparatus, but not for the two separately.

The establishment of a statistical correlation between system
and apparatus can make it possible to extract more energy out of
the system than can be done by the simple work cycles where the
external fields remain uncorrelated with the system. In Chapter
11 it is shown how it is possible to subdivide such an exchange
of energy between two quantum systems into work and heat.

Finally some of the relations between this work and other
approaches to the description of irreversibility is discussed in
Chapter 12.

CHAPTER 2. DYNAMICS AND WORK

In this chapter the dynamics of a finite system S subject to time-dependent external fields will be considered, and the work done by S on the fields is defined. For definiteness and generality a quantum description is chosen for S , but the basic concepts can equally well be defined for classical systems.

The microscopic observables of S are then the bounded self-adjoint operators in the separable Hilbert space H_S :

$$O_S = \{A \in B(H_S); A^+ = A\} \ ,$$

and the space of normal states is the set of density operators

$$E_S = \{\rho \in B(H_S); \rho \geq 0, \mathrm{Tr}\rho = 1\}.$$

S is acted upon by external fields, denoted by F, which can be controlled by the experimenter (e.g. the magnetic field in NMR experiments). The notation F will include boundary conditions and constraints. The fields will, in the first approximation, be considered to be macroscopic, i.e. not subject to fluctuations or back reaction from S. No distinction will be made between the system F and the set of values $\{f\}$ of the fields.

To start with it is also assumed that S is not coupled to the world outside $S + F$ in any way. For every value f of the field F there is then a (generally unbounded) Hamiltonian $H(f)$ on H_S which defines the evolution of S. The set of Hamiltonians which can be achieved by varying f

$$F = \{H(f); f \in F\}$$

is assumed to have the following properties

(1) If H_1, $H_2 \in F$, then $H_1 - H_2$ is bounded

(2) F is convex.

These conditions mean that any two elements H_1, $H_2 \in F$ can be connected by a norm continuous path

$$H(t) = H_1 + t(H_2 - H_1)/\tau , \quad t \in [0,\tau],$$

or by a corresponding sequence of small discrete steps (t integer). As an example take an N-particle system with given kinetic energies and interaction. The one-particle potentials may be of the form

$$V = V_0 + V_1(\delta)$$

where V_1 is a linear function of the external fields (as in e.g. the Stark and Zeeman effects). The set of achievable fields are limited in strength by the experimental apparatus: $|\delta| \leq$ constant, where $||$ denotes a suitable norm on the fields.

A time-dependent field $\delta(t)$ then generates a time-dependent Hamiltonian H(t) and a time evolution in E_S through

$$\frac{d}{dt} \rho(t) = - \frac{i}{\hbar}[H(t),\rho(t)]. \tag{2.1}$$

A formal solution is given by a family of unitary propagators U(s,t) (defined by the standard Dyson expansion [8]) through

$$\rho(t) = U(t,s)\rho(s)U(t,s)^+ \tag{2.2}$$

For bounded and piecewise strongly continuous functions H(t) the Dyson expansion is convergent and the solution is welldefined. For unbounded Hamiltonians the situation is much more complex. Some general existence theorems are given in [8-10]. These difficulties are not really essential in the present context, and they are most easily avoided altogether by restricting the time dependence of the Hamiltonian to piecewise constant functions.

It is convenient to allow instantaneous changes in the fields also for the reason that the thermodynamic functions (to be defined later) would otherwise be functions not only of the state of S but also of the state of F.

The unitary propagators are then welldefined and so are the dynamical maps on O_S:

$$T(s,t)[A] = U(t,s)^+ A U(t,s), \qquad A \in O_S$$

$$T(s,t) \cdot T(t,u) = T(s,u) .$$

Let T^* denote the dual maps on E_S given by (2.2). The general relation between the Heisenberg picture map T (on the observables) and the dual Schrödinger picture map T^* (on the states) is defined by

$$T^*[\rho](X) = \rho(T[X]),$$

for all $\rho \in E_S$, $X \in O_S$.

In dealing with the concept of work and defining the thermodynamic functions the crucial role is played by the cyclic changes in the fields.

Definition. A *work cycle* is a piecewise constant map from an interval of R to F which starts and ends at one and the same $H \in F$

$$\gamma = \{H:[s,t] \rightarrow F; H(s) = H(t) = H\}.$$

We also introduce the notation

$$D(\gamma) = [s,t] = \text{the } \textit{duration} \text{ of } \gamma$$

$O(\gamma) = H =$ the *origin* of γ.

The set of all work cycles defined by F is denoted by $\Gamma(F)$, those of given origin H by $\Gamma(F,H)$, and those of given duration D by $\Gamma(F,D)$. Two cycles are called *equivalent* $(\gamma_1 \simeq \gamma_2)$ if they differ only in a finite number of points (this includes the assumption $D(\gamma_1) = D(\gamma_2)$, but in general $O(\gamma_1) \neq O(\gamma_2)$). One can trivially extend a cycle γ to a cycle defined on an interval $D' \supset D(\gamma)$ by putting H(t) equal to $O(\gamma)$ in $D' \diagdown D(\gamma)$.

Two work cycles γ_1, γ_2 with common origin H and durations $[s_1,t_1]$ and $[s_2,t_2]$, respectively, where $t_1 \leq s_2$, have a time-ordered composition

$$(\gamma_1,\gamma_2) \to \gamma = \gamma_1 * \gamma_2$$

defined by

$$H(t) = H_1(t) \text{ for } t \in [s_1,t_1],$$

$$= H \qquad \text{for } t \in [t_1,s_2],$$

$$= H_2(t) \text{ for } t \in [s_2,t_2].$$

Conversely, a work cycle γ on $[s,t]$ can be decomposed, up to equivalence, into two cycles γ_1 on $[s,u]$ and γ_2 on $[u,t]$ for any choice of $u \in (s,t)$, such that $\gamma \simeq \gamma_1 * \gamma_2$.

A work cycle γ defines the time evolution of S during $D(\gamma) =$ $= [s,t]$ in the way described above. We write

$$T(\gamma) \equiv T(s,t)$$

Note that the dynamical maps are indifferent to finite changes in H(t) in a finite number of points, i.e.

$$\gamma_1 \simeq \gamma_2 \Rightarrow T(\gamma_1) = T(\gamma_2).$$

13

The $T(\gamma)$ form a semigroup of transformations under a limited operation of composition. If the cycles γ_1, γ_2 have a common origin and if $D(\gamma_1)$ and $D(\gamma_2)$ are contiguous intervals, then

$$D(\gamma_1 \star \gamma_2) = D(\gamma_1) \cup D(\gamma_2)$$

$$T(\gamma_1 \star \gamma_2) = T(\gamma_1) \cdot T(\gamma_2)$$

We call

$$T(F) = \{T(\gamma); \ \gamma \in \Gamma(F)\}$$

$$= \{T(\gamma); \ \gamma \in \Gamma(F,H)\}, \ \text{any} \ H \in F,$$

the *mobility semigroup* generated by F. This is one of the basic concepts of the present work, and it owes much to the *global mobility* of Mielnik [11] (see also Waniewski [12]). Similar notions exist in system theory, where the input variables correspond to F.

It will be necessary to consider also mobility semigroups associated with cycles restricted to a finite time interval, $[0,\tau]$ say, corresponding to the time the experimenter can keep S isolated from the external world. There is then a lack of time homogeneity which forces us to keep track of the time parameter t of a state ρ and apply only cycles γ with $D(\gamma) \subset [t,\tau]$ to ρ. This restriction will be assumed throughout the following wherever it is applicable.

The range of dynamical maps in T(F) clearly increases with the size of F (and with τ where appropriate). The greater the experimenter's control of the dynamics of S, the larger is the set of motions he can force S to perform during a given time interval, or the shorter the time needed to perform a given transformation. If F is large enough, e.g. if $F = 0_S$, then all unitary transformations will be in T(F), which then forms a group. This case corresponds to the idea of a reversible microscopic system.

It is more difficult to estimate the size of T(F) when F is small. The interesting case for thermodynamics is where T(F) is not a group and where it forms a very small part of the set of all unitary transformations of O_S . It will be argued in Chapters 6 and 10 that this lack of control of S can be seen as a source of irreversibility.

Note that even if F is small it may be possible to have an effective reversibility. In the spin-echo experiment [5] the free evolution (defined by some γ) where the sample magnetization decays can be reversed by applying a pulsed magnetic field (represented by γ') such that the initial state is approximately restored

$$T(\gamma')^* \cdot T(\gamma)^*[\rho] \simeq \rho .$$

In this case F is defined by two free parameters while S is practically infinite. For finite systems there is the phenomenom of recurrence which occurs if some $T(\gamma)$ has a discrete spectrum. Then for any $\rho \in E_S$ and $A \in O_S$ the sequence $\rho(T(\gamma)^n[A])$ is almost periodic, i.e. for every choice of $\varepsilon > 0$

$$|\rho(T(\gamma)^n[A]) - \rho(A)| < \varepsilon$$

holds for an infinite sequence of values of n. The problems associated with this type of reversibility will be treated in Chapter 6.b.

The mobility semigroup defines which states can be reached from a given initial state using the control of the system given by the fields. We say that a state μ is F-accessible from ρ if there is a $\gamma \in \Gamma(F)$ such that

$$\mu = T(\gamma)^*[\rho] .$$

The set of all states F-accessible from ρ is denoted by

$$\Omega(\rho,F) \equiv T(F)^*[\rho].$$

The following properties are crucial:

$$\Omega(\rho,F) \subseteq \Omega(\rho,F') \tag{2.3}$$

for $F \subseteq F'$, and

$$\mu \in \Omega(\rho,F) \Rightarrow \Omega(\mu,F) \subseteq \Omega(\rho,F) \tag{2.4}$$

The first relation is obvious, the second follows directly from the semigroup property of $T(F)$.

Definition. The *work* performed on the fields F during a work cycle γ by the system S, which is initially in the state ρ at time s, is

$$W(\gamma,\rho) = - \int_{D(\gamma)} \rho(u)[dH(u)] \tag{2.5}$$

where

$$\rho(u) = T(s,u)^*[\rho]$$

is the time evolution of the state defined by γ in $D(\gamma) = [s,t]$.

The expression (2.5) is immediately integrated to yield

$$W(\gamma,\rho) = (\rho(s) - \rho(t))[H] = \rho[\hat{W}(\gamma)] \tag{2.6}$$

where $H = O(\gamma)$ and the work observables

$$\hat{W}(\gamma) = H - T(\gamma)[H]$$

satisfy, for contiguous cycles,

$$\hat{W}(\gamma_1 * \gamma_2) = \hat{W}(\gamma_1) + T(\gamma_1)[\hat{W}(\gamma_2)].$$

Due to the assumptions on F they are all bounded for cycles of finite duration (which have a finite variation of $H(t)$), but they are generally not in F. Note that equivalent cycles of different origin give different values for the work as is obvious from (2.6).

The preceding formalisn can immediately be applied to classical Hamiltonian dynamical systems. With some care systems without
a regular Hamiltonian, such as a hard sphere gas in a variable 1-
particle potential, should be amenable to the same treatment.

Provided that the work can be defined, there is also a generalization to systems with non-Hamiltonian dynamics. Consider the
case where S is a subsystem of a large Hamiltonian system $S + R$
(i.e. S is an open system, see Appendix A.1 for a short introduction to this subject). If F is defined relative to $S + R$ in such
a way that

$$H_1, H_2 \in F \Rightarrow H_1 - H_2 \in \mathcal{O}_S \otimes I_R$$

then the formula (2.5) can be expressed completely in terms of
the partial state of S:

$$W(\gamma,\rho) = - \int \rho_S(u)[dH_S(u)] \qquad (2.7)$$

where ρ_S is defined by the equality

$$\rho_S[A] = \rho_{S+R}[A \otimes I_R], \text{ all } A \in \mathcal{O}_S.$$

The simplest model for the reduced dynamics of S is given by a
Markovian master equation where (2.1) is replaced by (A.12) of
Appendix A:

$$\frac{d}{dt} \rho(t) = L^*[\rho(t)] - \frac{i}{\hbar}[H(t),\rho(t)].$$

L is a generator of the form given in (A.7) which describes the
influence of R on the evolution of S (see also [13-15] for a background). The family of propagators obtained by integration are
completely positive maps and have the same composition rule as
before. In this case the formula for the work is not directly
integrable to an expression like (2.6) involving only the partial

17

state of S.

It is finally necesary to specify what type of experiment the
formalism describes and which quantities will be considered as
observable. At some instant ($t = 0$, say) S is in an equilibrium
state due to a preparation procedure in $t < 0$ (which will be described in Chapter 4). The action of time-dependent fields during
some time interval $[0,\tau]$ drives the system out of equilibrium. The
observed quantities are the ensemble averages of the work performed
in any subinterval of $[0,\tau]$, as well as the average energy transfer to the heat baths introduced in Chapter 4. The experiments
considered thus form a kind of generalized calorimetry. All these
mean values are predictable from the knowledge of the initial
state, and no additional information is obtained by the observation of S. In this way some intricate problems are avoided. First,
there is no need to introduce explicitly the measurement process
into the formalism. Secondly, the entropy will clearly be a function of the ensemble averages, thus avoiding the confusion which
sometimes enters when the entropy is allowed to be a fluctuating
quantity (as happens in some discussions of Boltzmann's H-theorem).
Thirdly, the problems of interpretation associated with the idea
of *Maxwell's demon*, which are inevitable in discussing measurements capable of discerning thermal fluctuations, can be saved to
the end. In Chapter 11 this and other aspects of the relation between measurements and entropy are discussed using the already
developed formalism.

A physical application where the experiments are of this type
is provided by NMR. The magnetization of the system is observed
through the dipole radiation. This radiation comes from the col-
lective dipole moment of all the spins, as it is coherent over the
sample. The back reaction on each individual spin is insignificant
and so is the energy loss. What one observes is a macroscopic non-
fluctuating quantity, and it is not necessary to introduce the

full quantum measurement theory. The quantum features of the system are made evident through its response to time-dependent magnetic fields.

It is interesting to note that already the measurement processes, which are necessary in order to define the quantum observables, break the time reversal symmetry of the formalism. So do the state preparing procedures of Chapter 4. Consequently, for a given initial state of the system, the work cycles can be chosen arbitrarily in the future relative to a time order which is given a priori by the relevant laboratory procedures. Thus the present formalism has nothing to say about the origin of *time's arrow*. There is a discussion of that subject in Davies [2] .

CHAPTER 3. INFORMATION ENTROPY

This chapter outlines the properties of the entropy function used in information theory. The Gibbs states, which satisfy the familiar variational principles with respect to this entropy, are also introduced.

3.a Entropy and relative entropy

The properties of the information (= statistical) entropy for quantum and classical systems have been surveyed recently by Wehrl [16] and Thirring[17]. Here the definitions and some properties which will be used further on are recalled. They will be formulated in the quantum case only. We will consider a quantum system S living in a Hilbert space H and use the notation $E_S = E(H)$ etc. where this is convenient.

To every density operator $\rho \in E(H)$ is associated the *information entropy* (I-entropy for short) defined by

$$S_I(\rho) = - \text{Tr}(\rho \ln \rho).$$

It is the most immediate generalization of Shannon's information measure in commutative probability theory to the quantum domain. It is chosen to be dimensionless by dropping Boltzmann's constant, which is generally included in physics literature. The most basic properties are as follows.

(1) *Positivity*: For all $\rho \in E(H)$

$$S_I(\rho) \geq 0 \ ,$$

where equality holds if and only if ρ is a pure (vector) state. In infinite-dimensional Hilbert spaces the value is often $+ \infty$.

(2) *Unitary invariance:* For all unitary operators U on H, all ρ

$$S_I(U^+\rho U) = S_I(\rho). \tag{3.1}$$

(3) *Concavity:* For all $\{\rho_i \in E(H), \lambda_i \geq 0; \sum_i \lambda_i = 1\}$

$$\sum_i \lambda_i S_I(\rho_i) \leq S_I(\sum_i \lambda_i \rho_i). \tag{3.2}$$

(4) *Subadditivity:* If $\rho \in E(H)$, where $H = H_1 \otimes H_2$ and the partial states ρ_1 and ρ_2 are defined by

$$\rho_1 = Tr_2\rho \in E(H_1), \qquad \rho_2 = Tr_1\rho \in E(H_2),$$

then

$$S_I(\rho) \leq S_I(\rho_1 \otimes \rho_2) = S_I(\rho_1) + S_I(\rho_2), \tag{3.3}$$

where equality holds if and only if $\rho = \rho_1 \otimes \rho_2$.

It is an interesting fact that (3.1), (3.3) and a very weak continuity condition suffice to characterize completely the function S_I [18]. However, (3.1) implies at once that S_I is not a suitable candidate for an entropy function which is intended to describe irreversibility in the closed system dynamics of Chapter 2.

In addition to S_I it is convenient to introduce a closely related function which gives an entropic distance between two states. The *relative entropy* is a function of two states which can be written in a formal way (a more careful definition can be found in [19])

$$S_I(\rho|\mu) = Tr(\rho\ln\rho - \rho\ln\mu).$$

Note that the order of ρ and μ in this notation is opposite to that of [16,17]. The relative entropy has the following properties.

(1) *Positivity:* For all $\rho, \mu \in E(H)$

$$S_I(\rho|\mu) \geq 0 \qquad\qquad\qquad (3.4)$$

where equality holds if and only if $\rho = \mu$. The value may be $+\infty$ even in finite-dimensional Hilbert spaces.

(2) *Unitary invariance*: For all unitary $U \in B(H)$ and all ρ,μ

$$S_I(U^+\rho U|U^+\mu U) = S_I(\rho|\mu).$$

(3) *H-theorem*: For any completely positive map T on $\mathcal{O}_S$ such that T[I] = I (which means that the dual map T^* preserves probability) and all $\rho,\mu \in E(H)$,

$$S_I(T^*[\rho]|T^*[\mu]) \leq S_I(\rho|\mu) \qquad\qquad (3.5)$$

See Appendix A.1 or [13] for a definition of complete positivity and an explanation of the importance of this concept for dynamical maps. A proof of (3.5) can be found in [20,21]. This inequality is actually a version of the strong subadditivity property of the I-entropy and the proof is based on the results of Lieb [22] and Lieb & Ruskai [23]. It applies to the evolutions generated by Markovian master equations of the form (A.10) and gives a quantum H-theorem for them [24,25](see Appendix A.2, especially (A.13)).

An interpretation can be given of the relative entropy as a measure of how easy it is to distinguish ρ from μ by observing the system. This can be seen in an intuitive way by considering the commutative analogue defined for two relatively absolutely continuous probability measures on a space X

$$S_I(\rho|\mu) = \int_X d\rho(x) \; \ln\left[\frac{d\rho}{d\mu}(x)\right]$$

where $\frac{d\rho}{d\mu}$ is the Radon - Nikodym derivative. If $\rho(A) \geq a\mu(A)$ for some $A \subset X$, $a > 0$, then

$$S_I(\rho|\mu) \geq \rho(A)(\ln a - 1 + a^{-1}).$$

22

If $\ln a \gg \rho(A)^{-1}$, then the observation of the event $x \in A$ (with probability $\rho(A)$) is a strong evidence in favour of the hypothesis that the system is in the state ρ rather than μ, and correspondingly $S_I(\rho|\mu)$ is large. On the other hand, if $\frac{d\rho}{d\mu} \leq b = O(1)$, then ρ does not deviate too much from μ except on sets of low ρ-probability, it is difficult to tell ρ apart from μ and correspondingly $S_I(\rho|\mu) \leq b$.

With this interpretation the H-theorem can be seen as a most general property of loss of information for the reduced dynamics of open systems described in Appendix A.1,2. The variation of $S_I(\rho(t))$ in time is in general not of a definite sign for open systems, but nevertheless equation (3.5) describes a loss of information in the sense that the evolution makes the states less distinguishable. The thermodynamic significance of this type of irreversibility will be discussed in Chapter 7. Some cases where the variation in S_I is of a definite sign are treated in section 3.c below.

3.b Gibbs states

For a Hamiltonian $H \in F$ with discrete spectrum the *Gibbs state* of inverse temperature $\beta = (k_B \theta)^{-1}$ is defined in the standard way:

$$\rho(\beta,H) = \exp(-\beta H - \ln Z)$$

$$Z(\beta,H) = \mathrm{Tr}[\exp(-\beta H)]$$

The I-entropy and energy of this state are

$$S(\beta,H) \equiv S_I(\rho(\beta,H)) = \beta E(\beta,H) + \ln Z(\beta,H)$$

$$E(\beta,H) \equiv \rho(\beta,H)[H] = -\frac{\partial}{\partial\beta} \ln Z(\beta,H).$$

From (3.4) follows that

$$S_I(\rho|\rho(\beta,H)) = S(\beta,H) - S_I(\rho) + \beta[\rho(H) - E(\beta,H)] \geq 0 \quad (3.6)$$

and hence the well-known variational principles of thermodynamics (also called *thermodynamic stability*)

$$S_I(\rho) = S(\beta,H) \Rightarrow \rho(H) \geq E(\beta,H),$$

$$\rho(H) = E(\beta,H) \Rightarrow S(\beta,H) \geq S_I(\rho).$$

We use the notation

$$G(\beta,F) = \{\rho(\beta,H); H \in F\}$$

$$G(F) = \{\rho(\beta,H); \beta \in (0,\infty), H \in F\}.$$

It is necessary to restrict the set F in order to have Gibbs states for every H ∈ F with the properties we expect of them. In order that the construction to be made in Chapters 6 and 10 shall be possible, it is necessary that the range $\{E(\beta,H); \beta \in (0,\infty)\}$ is large enough for every H ∈ F. The simplest condition to guarantee this is the following.

Assumption: Each H ∈ F has a discrete spectrum which is unbounded above and has a non-degenerate ground level $E_0(H) > -\infty$.

This hypothesis excludes e.g. finite spin systems, which have an upper bound on the energy spectrum. They are thus capable of having *negative temperature states*, where the higher energy levels are more populated than the lower ones. These systems can be allowed by less restrictive assumptions (see Chapter 6).

Under the assumption made above Z, E and S are well-defined functions of (β,H) and

$$\lim_{\beta \to \infty} E(\beta,H) = E_0(H),$$

$$\lim_{\beta \to 0} E(\beta,H) = \infty ,$$

$$\frac{\partial}{\partial \beta} E(\beta,H) = E(\beta,H)^2 - \rho(\beta,H)[H^2] \leq 0,$$

where the last equality holds if and only if $\beta = \infty$. Consequently, for each $H \in F$ and $\rho \in E(H)$, the equation

$$E(\beta,H) = E \equiv \rho(H) > E_0(H)$$

has a unique solution

$$\beta = \beta(E,H) > 0.$$

The relation

$$\frac{\partial}{\partial \beta} S(\beta,H) = \beta \frac{\partial}{\partial \beta} E(\beta,H)$$

provides a justification for identifying $S(\beta,H)$ (multiplied by k_B to obtain the physical dimension) with the thermodynamic entropy of the equilibrium state $\rho(\beta,H)$ of the system. It implies that

$$S(E,H) \equiv S(\beta(E,H),H) = . \int_{E_0(H)}^{E} du \ \beta(u,H) + S_0(H) \qquad (3.7)$$

$$S_0(H) \equiv \lim_{\beta \to \infty} S(\beta,H).$$

As $E_0(H)$ has been assumed to be non-degenerate it is consistent to put, as a normalization of the entropy,

$$S_0(H) = 0, \quad \text{all } H \in F \ .$$

In a formal sense this can be interpreted as the third law of thermodynamics, while the physical low temperature behaviour is determined by the density of states near the ground level. For simplicity this assumption, as well as the energy normalization

$$E_0(H) = 0, \quad \text{all } H \in F,$$

will be used below. The expression (3.7) then gives a range of values for the entropy (given H) for finite temperatures

$$0 < S(\beta,H) < \lim_{\beta \to 0} S(\beta,H).$$

It does not follow from the assumption above that the upper bound is ∞, although this is so for most physical systems.

Using the unitary invariance (3.1) of S_I and (3.6), a convenient formula is obtained for the work (2.5) in a cycle γ with $O(\gamma) = H$, $D(\gamma) = [s,t]$:

$$W(\gamma,\rho(s)) =$$

$$= \beta^{-1} \{S_I(\rho(s)|\rho(\beta,H)) - S_I(\rho(t)|\rho(\beta,H))\} \qquad (3.8)$$

This relation holds for all β. It immediately gives the following conclusions:

(1) If $\rho(s) = \rho(\beta,H)$ for some β, then $W(\gamma) \leq 0$ for all cycles γ of origin H. This is the property of *passivity*, i.e. Kelvin's form of the second law, which has been discussed in detail (mainly for infinite systems) by Pusz and Woronowicz [26].

(2) If $\rho(s) = \rho(\beta,H)$ and $W(\gamma) = 0$, then $\rho(t) = \rho(s)$. In order to create a state ρ different from the initial state, an amount of work equal to

$$|W| = \beta^{-1} S_I(\rho|\rho(\beta,H))$$

has to be performed on the system.

(3) If $\rho(\beta,H)$ can be reached from the initial state $\rho(s) = \rho$ for some choice of β and γ, then the maximal work which can be extracted from the system through a work cycle of origin H is

$$A(\rho;H) = \beta^{-1} S_I(\rho|\rho(\beta,H)) \qquad (3.9)$$

This expression is sometimes taken as a sufficient reason to identify the RHS with the available work in an arbitrary non-equilibrium state ρ of S in an environment specified by β [27].

26

In general, however, no Gibbs state is accessible from ρ using the Hamiltonian dynamics defined in Chapter 2, unless the system behaves in a trivial, reversible way. This point will be clear from the following chapters, especially the discussion in Chapter 10.

It will be necessary to consider also infinite systems in order to define the notion of a *heat bath*. As equilibrium states for infinite systems we choose KMS states [17]. The reason for this choice is the various stability properties satisfied by KMS states and which have been reviewed by Thirring [17] and Sewell [28]. These include the variational principles and passivity described above for finite systems, as well as properties which are relevant only for infinite systems, some of which will be mentioned in Chapter 4. For simplicity we will sometimes use the name Gibbs state also for KMS states, which can in fact be considered as thermodynamic limits of Gibbs states for finite subsystems.

3.c Entropy-increasing processes

Under the unitary time evolution of a closed finite system the I-entropy and the relative I-entropy are constant. This property makes it impossible to use them as measures of an intrinsic irreversibility of the dynamics. There has been numerous attempts to find an alternative entropy function while still staying in the information theory framework. They have been reviewed by Wehrl [16], Penrose [1], see also Goldstein and Penrose [29]. A monotonic increase of the new entropy function is usually achieved by discarding "irrelevant information" in some way. The most elementary way of doing this is by neglecting the correlations between subsystems of the observed system.

Consider a system S consisting of two subsystems $S = S_1 + S_2$ which interact during a finite time interval. The initial state of S is taken to be $\rho = \rho_1 \otimes \rho_2$. The final state is called ρ'.

With the notation used in (3.3), it follows from (3.1) and (3.3) that

$$S_I(\rho_1) + S_I(\rho_2) = S_I(\rho) =$$

$$= S_I(\rho') \leq S_I(\rho'_1) + S_I(\rho'_2) \tag{3.10}$$

If the correlations measured by

$$C_{12}(\rho') \equiv S_I(\rho'_1 \otimes \rho'_2) - S_I(\rho') \geq 0$$

can be discarded in a consistent way, and if the total I-entropy of S is redefined as the sum of the partial entropies of S_1 and S_2, then the I-entropy increases due to the interaction. This is the mechanism behind the *Stosszahlansatz* or *molecular chaos* hypothesis used in the derivation of Boltzmann's H-theorem. The difficulties of justifying these assumptions by arguing that the correlations will not influence the future evolution of S will be commented upon in Chapter 7.

A change in the value of the I-entropy of S can also happen as a result of perturbations due to the external world. In general the interaction of a system with a reservoir will not give a definite sign of the change, as the entropy change in the reservoir must also be taken into account. However, in some problems the perturbations can be modelled by a stochastic term in the Hamiltonian. This type of model has been treated by Primas [30] and Kossakowski [31] , among others. The resulting dynamics causes an increase in S_I. The dynamical maps are of the type

$$T^*[\rho] = \int_X d\mu(x) \ U(x)\rho U(x)^+$$

where U: $X \rightarrow B(H)$ is a unitary map - valued random variable and μ is a probability measure on the space X. Due to (3.1) and (3.2) any map of this form satisfies, for all $\rho \in E(H)$

$$S_I(T^*[\rho]) \geq S_I(\rho)$$

Under some "white noise" conditions on the stochastic part of the
Hamiltonian, a semigroup evolution of the type defined by (A.4)
and (A.7) is obtained. The generator is of the form (A.7) with

$$L_d^\star[\rho] = \varepsilon^2 \sum_i \; [[V_i, \rho], V_i], \qquad V_i^+ = V_i \in B(H), \qquad (3.11)$$

where ε measures the strength of the perturbation. From the inequality above it then follows that

$$S_I(\rho(s)) \leq S_I(\rho(t)) \;, \quad \text{for all } s < t, \qquad (3.12)$$

for the solutions of (A.7) with the generator (3.11). This is
actually a version of (3.10) where S_2 is a reservoir with $\beta = 0$
and consequently with $\Delta S_I = 0$ (see (3.13) below). In fact the generator (3.11) is that of an open system in contact with a reservoir of infinite temperature (see Appendix A.1).

In many cases the empirical content of the notion of a closed
system is the lack of a detectable exchange of energy with the
external world rather than an absolute lack of interaction. This
consideration leads to the concept of an *essentially isolated system* used by Tolman [32], which is particularly relevant in a
thermodynamic context where the energy exchange is in the focus.
Blatt [33] pointed out that for non-equilibrium states the change
in the I-entropy due to the perturbations can be much more significant than the change in energy. This idea will be discussed
in more detail in Chapter 8. Here we will consider the changes in
the I-entropy which are allowed when the energy exchange with
other systems is negligible.

Consider a finite system S interacting during a finite time
with a large but finite system R. The initial state of $S + R$ is
again taken to be $\rho = \rho_1 \otimes \rho_2$. Now ρ_2 is assumed to be a Gibbs
state corresponding to the unperturbed Hamiltonian H_2 of R:

$$\rho_2 = \rho(\beta, H_2).$$

From (3.10) we know that the I-entropies of the partial states of S and R satisfy

$$\Delta S_I^S + \Delta S_I^R \geq 0 \ ,$$

where $\Delta S_I^S \equiv S_I(\rho_1') - S_I(\rho_1)$, etc. But (3.6) says that

$$S_I(\rho_2'|\rho_2) = -\Delta S_I^R + \beta \Delta E^R \geq 0.$$

Hence the I-entropy increase in R always satisfies

$$\Delta S_I^R \leq \beta \Delta E^R. \tag{3.13}$$

Provided that the switching on/off of the interaction does not perform work on $S + R$ the energy defined by the unperturbed Hamiltonians is conserved

$$\Delta E^S + \Delta E^R = 0$$

and hence

$$\Delta S_I^S \geq -\Delta S_I^R \geq -\beta \Delta E^R = \beta \Delta E^S.$$

Consequently, if only those processes are considered where we can observe that $\Delta E^S = 0$, then it can be concluded that the I-entropy can not decrease as a result of the interaction with R:

$$\Delta S_I^S \geq 0 \ . \tag{3.14}$$

On the other hand, if S is in a Gibbs state $\rho(\beta, H_1)$, then as in (3.13)

$$\Delta S_I^S \leq \beta \Delta E^S. \tag{3.15}$$

Hence, in order to have $\Delta S_I^S > 0$ in this case, an energy transfer from R to S is necessary. This can be seen as an expression of a stability property of Gibbs states under the randomizing influence of small external perturbations.

For non-equilibrium states (3.15) need not hold, and this fact will be important in Chapter 8. A rigorous treatment seems difficult in this case, but a rough estimate shows that ΔS_I^S can be very large indeed compared to (3.15). Consider for simplicity states of S specified by distributions over the energy levels e_k of H, i.e. of the form

$$\rho = \textstyle\sum_k p_k P_k \;,$$

where P_k is the one-dimensional eigenprojection of e_k and $\sum p_k = 1$. Let S be perturbed by R in such a way that only levels e_k within energy intervals Δ_i (where $\cup\, \Delta_i = (0,\infty)$) of length $|\Delta_i| = \delta$ are scrambled, leading to a maximal energy transfer $\Delta E = \delta$. Denote the number of states in Δ_i by $n(i)$. Then the maximal value of ΔS_I^S is calculated as follows. The minimal I-entropy is obtained with only one occupied level per Δ_i (probability $p(i)$), the maximal one with all levels equally occupied ($p_k = p(i)/n(i)$ for $e_k \in \Delta_i$). The difference is then

$$\Delta S_I^S = \textstyle\sum_i p(i) \ln n(i).$$

Consequently we obtain the estimate (for this class of states)

$$\Delta S_{I,max}^S \simeq \ln(d\Delta E^S) \;, \tag{3.16}$$

where d is the number of states per unit energy interval. d is expected typically to increase exponentially with the size of S, leading to a very high upper bound (3.16) for macroscopic systems.

CHAPTER 4. HEAT BATHS

A set of idealized quasifree heat baths are introduced in order to achieve the following goals:

(1) To have a method of preparing the observed system in any one of a set of welldefined, reproducible initial states.

(2) To have in the formalism the class of reversible isentropic processes of classical thermodynamics and the notion of temperature.

Basic to any analysis of a dynamic system is the ability of the experimenter to prepare the system in a welldefined initial state. Indeed the notion of a state in statistical theories is most often taken to represent an ensemble of systems, all prepared in an identical way. In dealing with microscopic systems it is often possible to prepare them in pure states. In a thermodynamic formalism it is natural to restrict the preparation procedures to those which consist of the ageing of the system while it is in contact with a heat bath of given temperature and with prescribed values of the external fields F. The prepared state is then assumed to be the equilibrium (Gibbs) state defined by the given external parameters. This means that the memory of the distant past has been obliterated, i.e. no statistical correlations exist between the past history of the system and the outcome of future observations on it. This is a highly non-trivial assumption as one can see from the existence of longlived metastable states which retain such a memory of the past history (e.g. diamond structure of carbon at normal temperature and pressure).

Clearly one can not hope to prove that a general infinite

system will relax to equilibrium if left alone. For finite closed systems even the concept of relaxation to equilibrium is as yet undefined in the formalism. In fact, the ultimate objective of this work is to serve as a conceptual basis for the description of such relaxation processes. Thus, the need to have a state preparing procedure in order to construct the formalism leads to some fundamental problems of an almost philosophical nature.

The act of preparing the initial state is an irreversible process in an intuitive understanding of that word, as the memory of the distant past is wiped out. The introduction of a priori irreversible preparing procedures in a formalism which has some ambition to explain irreversibility on the basis of reversible microscopic dynamics may seem to constitute a vicious circle. A similar argument, implying that the introduction of a priori probabilities (which corresponds to state preparation) can not be based on reversible microdynamics, led Krylov [34] to the conclusion that statistical physics can not be founded on classical or quantum mechanics.

As was already noted in Chapter 2, the measurement processes defining the algebra of observables of a quantum system do unavoidably introduce irreversibility in the formalism. However, it seems desirable to reduce the a priori introduction of irreversible processes as much as possible. For this reason the state preparing procedures will be defined by a class of highly idealized heat baths which can be specified in a microscopic way. They also have a sufficiently simple dynamics for a rigorous mathematical treatment.

Obviously the heat baths must be infinite systems in order to be able to absorb energy transferred from other systems without changing their equilibrium properties. For the same reason they must have a *return to equilibrium* property. Local disturb-

ances should relax to the equilibrium (KMS) state of the heat
bath. As the purpose of the heat baths is to force the observed
system to a Gibbs state, we must know how to couple them to a
small system S and find the relaxation properties of S. It seems
that the only class of systems where our knowledge about the relaxation properties is sufficient to do this consists of the infinite quasifree boson or fermion systems. Their return to equilibrium is of a rather trivial kind as local perturbations fly
off to infinity without equipartition between different degrees
of freedom [35,36]. The occurrence of Bose condensation makes
fermion heat baths the most convenient choice. We use the notation
$R(\beta)$ for such a heat bath of inverse temperature β.

The quasifree model of heat baths has been used often in the
theory of open systems. The simple expressions for the higher order correlation functions for quasifree systems and an assumption
of a decay faster than $(1 + |t|)^{-\delta}$ for some $\delta > 0$ for the two-
time correlation function imply that the conditions for Davies'
results on the weak coupling limit (WCL) are fulfilled. Davies
could show that in the WCL the reduced dynamics of an open system
S interacting with a reservoir R is governed by a master equation
of the type (A.7) (with time-independent Hamiltonian) provided
that the correlation functions of R satisfy certain assumptions
on their decay [37-40]. The relaxation takes place on a time scale
which is renormalized by the S-R coupling parameter. Note that
the case of a time-dependent Hamiltonian is more complex and does
not give an evolution of the form (A.10)-(A.12) unless the variation in H takes place in the rescaled time parameter [41].

The Gibbs state for S defined by the unperturbed Hamiltonian
of S and the temperature of R is a stationary state for the WCL
of the reduced dynamics of S (this is equation (A.8)). If the S-R
interaction satisfies an irreducibility condition (which is a generic property, i.e. true for "most" interactions), then every

initial state will relax to this unique stationary state [42].
The KMS states of R can actually be characterized among the stationary states by the *reservoir stability* property: For every
finite S (with given Hamiltonian) there is precisely one state of
S which remains stationary for the WCL of the reduced dynamics
under a sufficiently large class of S-R interactions (and this is
the Gibbs state of course) [43,44]. An infinite reservoir in a
KMS state is thus the only state preparing device which is available when the microscopic details of the S-R interaction can not
be controlled in a reproducible way.

Of course the KMS states satisfy a large number of other stability properties which are essential for their interpretation as
equilibrium states [28]. On of them is the *dynamical stability*,
which says that a small, local and time-independent perturbation
of the dynamics will result in a time evolution which has a stationary state "near" the unperturbed KMS state [45]. The exact
reduced dynamics of S will clearly have the same property. Models
of the type (A.7) will have it if the generator L is non-degenerate in the sense indicated above of having a unique stationary
state, i.e. satisfying $L^*[\rho] = 0$. Indeed the stationary state is
in general unique for sufficiently small perturbations under the
condition that there is a unique equilibrium state for the unperturbed dynamics. The exceptions are the cases where the set of
equilibrium states, considered as a function of the perturbation
parameter ε, has a bifurcation point precisely at $\varepsilon = 0$.

The corresponding stability property when S + R is subjected
to a random perturbation due to the rest of the world (called X
for short) is more complex. The unique equilibrium state of S + R
would be determined by the temperature of X (if defined), a process which corresponds to the "heat death" of our part of the universe. This is actually not the interesting aspect here. Instead
we must consider the situation where the coupling to X is very

weak and the total energy exchange is unimportant. It is thus necessary to use a limiting procedure where the equilibrium state of $S + R$ is dominated by the initial state of R. This can be done by coupling S weakly to R and X with a suitable ratio of the coupling parameters, which we call λ_R and λ_X, respectively.

In the WCL the explicit formula for the dissipative part L_d of the generator in (A.7) and the assumption that R and X are statistically independent give L_d as the sum of two contributions from R and X , respectively [24]:

$$L_d(\varepsilon) = L_R + \varepsilon^2 L_X ,$$

where $\varepsilon = \lambda_X/\lambda_R$. Let $\varepsilon \ll 1$ be fixed and let $\lambda_R \to 0$ in the WCL. In the generic case, if L_R has a unique stationary state for the evolution unperturbed by X, then so has $L_d(\varepsilon)$ for sufficiently small ε, and the new stationary state will be near the unperturbed one. Thus, if the experimenter can control the coupling λ_R to a given heat bath, then he can use it to dominate the uncontrolled perturbations from X provided that there is an upper bound on their strength. It is assumed that this strength of the perturbation is so small that the WCL is still relevant.

It may be of interest to recall an experimental realization of this state preparing procedure where the strength of the $S - R$ interaction can be controlled. In NMR experiments in solids the initial polarization of the sample is obtained when the spin system relaxes to equilibrium in a strong magnetic field. The temperature is defined by the crystal lattice and a suitable relaxation rate (defined by the spin-lattice coupling) is obtained through doping the crystal with paramagnetic impurities. In this way the time needed for the preparation of the ·initial state may be reduced to a few minutes even for very low temperatures and high magnetic fields. At the same time the spin system is practically closed during the much shorter time needed for the experiments on the spin-spin relaxation.

The model of the heat baths chosen here does perhaps need some justification besides that of being practically the only one available with the present knowledge of the dynamics of infinite systems. First, note that in classical thermodynamics all heat baths of a given temperature are interchangable, i.e. we use only some universal property of the system. Thus it can be argued that it is sufficient to have one microscopic model with this property even if it is rather artificial. Secondly, the notion of a heat bath is a gross idealization. We can not expect general infinite systems to behave exactly as this idealization would have us believe. In fact, judging from computer experiments it seems that even infinite classical systems will show a behaviour corresponding to relaxation to equilibrium only if the total energy is larger than a certain *stochastic transition threshold* [46]. Below this threshold the motion of the system is quasiperiodic, with every initial non-equilibrium state recurring indefinitely, and consequently it can not perform exactly as an ideal heat bath. Quantum systems will not have better properties in this respect. Of course, a quasifree fermion system has no such threshold and behaves as an ideal heat bath for every value of the temperature.

The mathematical rigour achieved through the use of the WCL is offset by the difficulty of a physical interpretation of the time rescaling. In the present formalism there are time-dependent external fields which must be allowed to change on a time scale given by the intrinsic dynamics of the small system S. It will be obvious from the considerations of Chapters 6 and 7 that this time dependence can not be rescaled. The only way out seems to be to accept two different types of processes with two separate time scales. First, there are work cycles acting on a closed system S with a natural time scale given by the dynamics of S. Secondly, there are processes where S is in contact with a heat bath (or several heat baths of different temperatures in succession) with a time parameter rescaled according to the weak coupling limit. Here we have both the state preparing procedures and the reversible processes of thermodynamics where the external field changes slowly compared to the relaxation processes induced by the heat baths.

It is now possible to specify the set of states which can be prepared through the processes defined above. The Gibbs states $G(F)$ are prepared by the action of one of the reservoirs while $H \in F$ is constant. The set of states which are F-accessible from $G(F)$ using the Hamiltonian dynamics defined by F (i.e. through the mobility semigroup $T(F)$) is

$$E_0(F) = \{\Omega(\rho,F); \ \rho \in G(F)\}.$$

It is mathematically convenient, though not always physically relevant, to have a convex state space for statistical models. Hence, we can define the set of *reproducible states* to be

$$E(F) = \text{convex hull of } E_0(F).$$

This set contains all the experimentally available non-equilibrium states, and it is for these states that we have to define the entropy functions.

CHAPTER 5. REVERSIBLE PROCESSES

Using the idealized heat baths and the WCL of Chapter 4, we can now introduce the class of reversible processes defined by F and which constitute classical thermodynamics. The finite system S can be coupled weakly to any one of the set of reservoirs $R(\beta)$, $\beta \in (0,\infty)$. Let S be prepared in the initial state $\rho(\beta,H)$ through the action of $R(\beta)$. Apply a work cycle $\gamma \in \Gamma(F,H)$ to S, which is then again coupled to $R(\beta)$ to recover the initial state. The total increase in the I-entropy of $R(\beta)$ is then given by the increase in the energy

$$\Delta S_I^R = - \beta W(\gamma,\rho(\beta,H)) = \beta \Delta E^R. \qquad (5.1)$$

Note that $W \leq 0$ due to the passivity of the initial state with respect to cycles of origin H.

The standard formula (5.1) merits some comments regarding the use of the I-entropy and the origin of irreversibility. Applying the formula (3.8) to $S + R$ as if this were a finite system we obtain

$$\beta W = - S_I(\rho_1|\rho_0)$$

where ρ_0 (the initial equilibrium state) and ρ_1 (the final state) are now considered as states of the finite system $S + R$. It is then possible to define an I-entropy increase in $S + R$ in the following way. By the assumption made on F in Chapter 3.b there is a unique $\beta' \leq \beta$ such that

$$E(\beta',H) = \rho_1(H) = \rho_0(H) - W.$$

Put $\rho_1' = \rho(\beta',H)$. From the identity

$$\beta W = S_I(\rho_0) - S_I(\rho_1') - S_I(\rho_1'|\rho_0),$$

and a similar one with β and β' exchanged it follows that

$$S_I(\rho_1'|\rho_0) \leq S_I(\rho_1'|\rho_0) + S_I(\rho_0|\rho_1') \leq (\beta' - \beta)W.$$

If, in the *thermodynamic limit* where the size of R goes to ∞, the heat capacity of $S + R$ goes to ∞ while W is fixed, then we obtain $\beta' = \beta$ and consequently

$$\lim_{|R| \to \infty} S_I(\rho_1'|\rho_0) = 0 \qquad\qquad (5.2)$$

Hence, in the thermodynamic limit

$$- \beta W = S_I(\rho_1|\rho_1') \equiv \Delta S_I^{S+R} \geq 0 \qquad\qquad (5.3)$$

where, for every finite approximation of this limit,

$$S_I(\rho_1) = S_I(\rho_0)$$

$$S_I(\rho_1|\rho_1') = S_I(\rho_1') - S_I(\rho_0).$$

Why the choice (5.3) rather than "ΔS_I" $= S_I(\rho_1) - S_I(\rho_0) = 0$? The answer lies in the transition to a quasilocal description of $S + R$ in the thermodynamic limit [17]. Let the symbol Λ denote finite parts of $S + R$. The local state ρ_Λ is then defined as a partial state for the subsystem Λ through

$$\rho_\Lambda(A) = \rho(A)$$

for all local observables belonging to Λ. If there is local relaxation back to equilibrium, then for any Λ

$$\rho_{1,\Lambda} = \rho_{1,\Lambda}' = \rho_{0,\Lambda}$$

holds in the limit $|R| \to \infty$, and the part of the information in ρ_1 which corresponds to ΔS_I^{S+R} disappears into global correlations. By definition these can not be measured in any finite Λ and they

40

do not influence the future evolution in the quasilocal model. We can display these correlations by using the identity

$$S_I(\rho_1|\rho_0) = S_I(\rho_1|\rho_1') + S_I(\rho_1'|\rho_0).$$

Due to (5.2) ρ_1' is the state of the finite approximation which corresponds most closely to the final quasilocal equilibrium state of $S + R$.

$$\Delta S_I^{S+R} = S_I(\rho_1|\rho_1') = [S_I(\rho_1|\rho_0)]_\infty$$

is then interpreted as the contribution from the global correlations present in ρ_1.

The loss of information due to the quasilocal description is a limiting case of the entropy increase due to the discarding of correlations described in equation (3.10). The correlations between S and R defined by the equilibrium states for $S + R$ disappear in the WCL, i.e.

$$\Delta S_I^{S+R} = \Delta S_I^S + \Delta S_I^R.$$

If S returns to the initial state in a cyclic process, then $\Delta S_I^S = 0$, and consequently (5.1) holds.

Now consider a special work cycle acting on S

$$H(t) = H_0 \text{ for } t \le 0 \text{ and } t > \tau,$$

$$= H_1 \text{ for } 0 < t \le \tau.$$

Let the initial state of S be $\rho_0 = \rho(\beta, H_0)$ and let S be weakly coupled to $R(\beta)$ during $(0,\tau)$, where τ is large compared to the relaxation time in the WCL. Then

$$\rho(\tau) \simeq \rho(\beta, H_1) \equiv \rho_1 \,,$$

and the work performed by S is

$$W \simeq (\rho_1 - \rho_0)[\Delta H], \qquad \Delta H \equiv H_1 - H_0.$$

Replace this cycle by N similar cycles with small steps $\varepsilon\Delta H$ in the Hamiltonian and put $\varepsilon = N^{-1}$. The work performed in the total cyclic process is then

$$W \simeq \sum_{k=0}^{N-1} \{\rho(\beta, H_0 + (k+1)\varepsilon\Delta H) - \rho(\beta, H_0 + k\varepsilon\Delta H)\}[\varepsilon\Delta H]$$

$$= \varepsilon(\rho_1 - \rho_0)[\Delta H].$$

Hence $\lim_{\varepsilon \to 0} W = 0$ and consequently, by (5.1):

$$\lim_{\varepsilon \to 0} \Delta S_I^R = 0.$$

From this relation and (3.10) follows that the total I-entropy is constant for any part of the limiting process:

$$\Delta S_I^S + \Delta S_I^R = 0.$$

The conclusion is then that the process of changing H while S is weakly coupled to a heat bath becomes reversible in the limit where the rate of change is much slower than the relaxation induced by the heat bath.

Another type of reversible process involves changing the temperature of S. Let S be in an initial state $\rho_0 = \rho(\beta_0, H)$. S is brought into contact with $R(\beta_1)$ and allowed to relax to the state $\rho(\beta_1, H)$, then again separated from $R(\beta_1)$ and brought back to the state ρ_0 through the action of $R(\beta_0)$. The total entropy increase in the heat baths is

$$\Delta S_I^R = \Delta\beta[E(\beta_0, H) - E(\beta_1, H)]$$

where $\Delta\beta = \beta_1 - \beta_0$. Again this process is replaced by N similar cycles involving N + 1 heat baths with temperature steps $\Delta\beta/N$. The total entropy increase is then

$$\Delta S_I^R = \sum_{k=0}^{N-1} N^{-1} \Delta\beta [E(\beta_0 + k\Delta\beta, H) - E(\beta_0 + (k+1)\Delta\beta, H)]$$

$$= N^{-1} \Delta\beta [E(\beta_0, H) - E(\beta_1, H)].$$

Hence $\lim_{N \to \infty} \Delta S_I^R = 0$, and again reversibility is obtained in the limit of infinitely slow change.

Combining the two types of processes defined above gives a set of reversible processes where the total I-entropy is conserved. The cyclic ones correspond to Carnot cycles in classical thermodynamics. A special class are the essentially adiabatic processes used by Tolman [32] where the net average energy exchange with every heat bath is zero (S is essentially isolated in the terminology of Tolman). The I-entropy of S is then constant, but note that the corresponding dynamical maps are non-unitary in general. In fact the state of S is always a Gibbs state in such a process and the spectrum of the state need not be conserved. Hence a unitary evolution can not preserve this canonical form of the state even in the *adiabatic limit* of a infinitely slow variation in the Hamiltonian or the temperature. The advantage of having these non-unitary transformations in the formalism includes the possibility of characterizing the equilibrium states through the passivity property (this will be done in Chapter 10).

A final remark: For infinite systems it holds under some restrictions on the dynamics that the invariance of the state under infinitely slow cyclic local perturbations characterize the KMS states [41]. This is yet another reason to associate the KMS states with thermodynamic equilibrium.

CHAPTER 6. CLOSED FINITE SYSTEMS

The object of this chapter is to discuss to what extent irreversibility can be defined for a finite quantum system with Hamiltonian dynamics. Thus we consider again the situation described in Chapter 2. A closed finite system S is given with a set F of Hamiltonians and the corresponding set $\Gamma(F)$ of work cycles.

6.a Available work

The time evolution defined by any $\gamma \in \Gamma(F)$ satisfies the basic relation (2.4), which we can write as

$$\Omega(\rho(t),F) \subseteq \Omega(\rho(s),F) \tag{6.1}$$

for all $s,t \in D(\gamma)$, $s \leq t$. In fact, from the definition follows that $\rho(t) \in \Omega(\rho(s),F)$.

If F is large enough for the mobility semigroup $T(F)$ to contain all unitary transformations, then for every $\gamma \in \Gamma(F)$ with $D(\gamma) = [s,t]$ there is a $\gamma' \in \Gamma(F)$ with $D(\gamma') \subset [t,\infty)$ such that $T(\gamma') = T(\gamma)^{-1}$. Hence

$$\rho(s) = T(\gamma')^*[\rho(t)],$$

$$\Omega(\rho(s),F) \subseteq \Omega(\rho(t),F).$$

Consequently equality holds in (6.1) for all (s,t) in this case, which represents a reversible system. A strict inequality in (6.1) is taken as a definition of *irreversibility* in the dynamics defined by F. This irreversibility will now be expressed in a more thermodynamic fashion in terms of the concepts of energy and work.

44

For every choice of $H \in F$ there will be a greatest lower bound on the energy of the states which are accessible from a given state ρ

$$Q(\rho;F,H) = \inf\{\mu(H); \mu \in \Omega(\rho,F)\}.$$

If the Hamiltonians in F are normalized to be non-negative, then

$$0 \leq Q(\rho;F,H) \leq \rho(H).$$

From (2.3) follows that

$$Q(\rho;F',H) \leq Q(\rho;F,H) \tag{6.2}$$

for $H \in F \subset F'$, and from (6.1) that

$$Q(\rho(s);F,H) \leq Q(\rho(t);F,H) \tag{6.3}$$

for $s \leq t$. For every $\gamma \in \Gamma(F,H)$

$$Q(\rho;F,H) \leq \rho(H - \hat{W}(\gamma))$$

due to (2.6) and in fact

$$Q(\rho;F,H) = \inf\{\rho(H - \hat{W}(\gamma)); \gamma \in \Gamma(F,H)\} \tag{6.4}$$

From this expression follows directly the concavity of Q:

$$\sum_i \lambda_i Q(\rho_i;F,H) \leq Q(\sum_i \lambda_i \rho_i;F,H) \tag{6.5}$$

for all $\{\rho_i \in E_S, \lambda_i \geq 0; \sum_i \lambda_i = 1\}$. From the passivity of the Gibbs states and (6.4) follows that

$$Q(\rho;F,H) = \rho(H) \tag{6.6}$$

for $\rho = \rho(\beta,H)$ or a convex combination of such states with different values of β. Clearly any state satisfying (6.6) is passive with respect to cycles in $\Gamma(F,H)$

From (6.4) follows that the maximal work we can get out of the system S using work cycles in $\Gamma(F,H)$, if it is initially in the state ρ, is given by the *availibility function*

$$A(\rho;F,H) = \rho(H) - Q(\rho;F,H). \qquad (6.7)$$

There must be a sequence $\{\gamma_\alpha \in \Gamma(F,H)\}$ such that

$$A(\rho;F,H) = \lim_{\alpha \to \infty} W(\gamma_\alpha,\rho),$$

but there need not be a proper limiting cycle $\hat{\gamma} \in \Gamma(F,H)$ such that

$$A(\rho;F,H) = W(\hat{\gamma},\rho). \qquad (6.8)$$

Instead we can consider the sequence $\{\gamma_\alpha\}$ to define an *optimal work cycle* in a generalized sense. The relation (6.8) is then to be read as: For every $\varepsilon > 0$ there is a $\gamma \in \Gamma(F,H)$ such that

$$W(\gamma,\rho) \leq A(\rho;F,H) \leq W(\gamma,\rho) + \varepsilon.$$

The relation (6.3) says that the part of the energy of the system which can not be recovered as work does not decrease in time. The evolution is considered to be irreversible in the thermodynamic sense if this non-recoverable energy Q does actually increase. Observe that if $\rho(s) = \rho(\beta,H)$, then equality in (6.3) implies that $\rho(t) = \rho(s)$ and hence equality holds in (6.1). For the evolution defined by an optimal work cycle equality holds in (6.3). In fact for every $\gamma \in \Gamma(F,H)$, $D(\gamma) \in [s,\infty)$ we can write

$$\gamma = \gamma_1 * \gamma_2$$

$$D(\gamma_1) = [s,t], \qquad D(\gamma_2) \subset [t,\infty),$$

The work performed by S in the two cycles is

$$W_1 = W(\gamma_1,\rho(s)) = (\rho(s) - \rho(t))[H],$$

$$W_2 = W(\gamma_2,\rho(t)) \leq \rho(t)[H] - Q(\rho(t);F,H).$$

Then, by (6.3),

$$W(\gamma,\rho(s)) = W_1 + W_2 \leq \rho(s)[H] - Q(\rho(t);F,H)$$

$$\leq \rho(s)[H] - Q(\rho(s);F,H).$$

(6.8) then gives equality in (6.3) for $\gamma = \hat{\gamma}$.

The properties (6.1) and (6.3) have their origin in the breaking of time reversal symmetry inherent in the definition of Ω and Q. The state preparing procedure combined with the definition of work cycles specifies a "past" and a "future" as pointed out in Chapter 2. Hence Loschmidt's reversibility paradox does not apply to the present formalism. The recurrence paradox of Zermelo is not so easy to dispose of for a finite quantum system.

6.b Recurrences

For classical dynamical systems with the *mixing* property the Poincaré recurrence theorem does not really present a difficulty, as reproducible states represented by continuous phase space densities will not show a return to the initial state [48]. An example of a Hamiltonian system with this property is provided by the geodesic motion on a compact Riemannian manifold of negative curvature [49]. In general, however, classical Hamiltonian systems will have regular regions in phase space where the motion is not mixing.

For a quantum system in a bounded region in space with a discrete energy spectrum mixing can not occur. Instead the motion is almost periodic and recurrences (returns near the initial state) take place for all initial states (see e.g. § 1.1 of [17]). This conclusion applies also in the present context of time-dependent Hamiltonians. In fact, the recurrence need not be spontaneous but can be facilitated through the choice of cycles which are optimal in this respect. As the example of the spin echo shows, the return to the initial state may be made to take place after a quite short time for some macroscopic systems.

The existence of recurrences, spontaneous or induced by the
fields, poses a severe conceptual problem in the definition of
irreversibility. If the system is prepared at $t = 0$ in an equilibrium state ρ, and if it can be made to return to this state at
some future time using the unitary maps of $T(F)$, in spite of the
external forces which are applied to it in any finite interval
$(0,t)$, then it obviously follows from (6.6) that

$$Q(\rho(t);F,H) = \rho(H)$$

for all $t > 0$ and H corresponding to $\rho = \rho(\beta,H)$. Consequently
equality holds in (6.3) and in (6.1) as well. The same is true if
the motion is almost periodic but the recurrence is only approximate for any finite time, at least if H is bounded. If H is unbounded the situation is not so obvious, but we will not examine
this detail.

Already Boltzmann recognized that the recurrence periods of
large systems are likely to be of such magnitude that the motion
is "practically" irreversible. Unfortunately the distinction between practical and impractical experimental procedures is vague
and a function of time as well. This is especially evident in the
present approach, and the spin echo experiment again provides an
example. The conventional way of repressing the recurrences involves moving them to $t = \infty$ by taking the thermodynamic limit or
by the addition of external noise (see Chapters 7-9). For classical
systems it is sufficient to make the dynamics irregular enough to
have the mixing property, as we already noted.

There is, however, a strong appeal in having non-trivial
measures of irreversibility also for closed, finite, regular systems, such that the thermodynamic limit or disturbances due to the
external world will not introduce qualitatively new features. This

48

approach seems necessary if we want to treat relaxation processes in small systems e.g. internal relaxation in molecules. A sketch of how this may be done will now be given. The basic idea is that for the type of experiments considered here the time homogeneity of the description is only an approximation. There must always be a specified (but not unique) time scale for them to tell us which recurrences to take into account.

Consider the time needed to extract a given amount of work from S in a given initial state. Let ρ be the state of S at time t and let F, H be fixed and left out of the notation for the moment. Define, for $Q(\rho) \leq E \leq \rho(H)$

$$\tau(\rho;E) = \inf\{\tau; Q(\rho;\tau) \leq E\}$$

$$Q(\rho;\tau) = \inf\{\rho(H - \hat{W}(\gamma)); \gamma \in \Gamma , D(\gamma) = [t,\tau]\}$$

This means that the time needed to extract the work $\rho(H)$ - E from S is $\tau(\rho;E)$ - t. For any motion defined by a $\gamma \in \Gamma$ it holds that

$$\tau(\rho(s);E) \leq \tau(\rho(t);E), \quad \text{all } s \leq t, \tag{6.9}$$

by an argument similar to that leading to (6.1). There is now no trivial equality for almost periodic systems. It may well happen that

$$\tau(\rho(t);E) - \tau(\rho(s);E) \gg t - s.$$

If this is the case, then the evolution in (s,t) has made it much more difficult to extract the given amount of work from the system. I consider this effect to be the origin and nature of irreversibility in finite systems.

The τ-functions are clearly related to the periods of recurrences. If the system can be made to return to the initial equilibrium state ρ after a time τ , then

$$\tau(\rho(t);E) \leq \tau$$

for $E = \rho(H)$. The practical possibility of utilizing the recurrences must depend on the order of magnitude of τ. In order to incorporate such vague notions of what is practical into the formalism, we must define them in terms of the available experimental procedures, i.e. the work cycles in our case.

It is obvious that the problem of extracting a maximal amount of work from S becomes meaningless when S can exchange significant amounts of energy with the rest of the external world. In Chapter 8 it is in fact argued that the perturbations due to the external world can be important also when the energy exchange is small, and that the isolation time can be an intrinsic property of S. The fact that the isolation of S can only be maintained for a limited time forces the conclusion that the problem is fundamentally non-homogenous in time. Assume that after the preparation of the initial state of S in $t < 0$ (where the heat baths dominate other external perturbations) the system can be considered closed to a given degree of accuracy during the finite interval $(0,\tau)$. This means for instance that the I-entropy is approximately constant in this interval.

In order to define thermodynamic functions characteristic of S itself, it is necessary to use work cycles of duration $(0,\tau)$ only (F,H are still fixed)

$$\Gamma(\tau) = \{\gamma \in \Gamma(F,H); D(\gamma) \subseteq [0,\tau]\}.$$

Still using the Hamiltonian dynamics of S, the definitions of Ω, Q, A, etc. (now functions of τ) can be rephrased in terms of this restricted set of work cycles. The properties (6.2) - (6.6) are still valid. In addition there is the evident inequality

$$Q(\rho;\tau') \leq Q(\rho;\tau), \qquad \text{all } \tau \leq \tau'. \tag{6.10}$$

Given constants δ,ε we can consider a value $\tau \gg \delta$ for which

$$Q(\rho;\tau + \delta) \leq Q(\rho;\tau) - \varepsilon \qquad (\delta,\varepsilon > 0)$$

as a period of partial recurrence for the system, given the initial state ρ at $t = 0$.

The inequality (6.3), which is no longer a trivial equality for almost periodic systems, provides a set of measures of the irreversibility in the evolution from $\rho(s)$ to $\rho(t)$ (one for each choice of F,H,τ). If F is so large that any unitary transformation in H_S may be generated in a time short compared to τ, then the system will still be seen as essentially reversible. As noted above an almost periodic system is likely to satisfy the recurrence property

$$\lim_{\tau \to \infty} Q(\rho(t);\tau) = Q(\rho(0);\infty), \qquad (6.11)$$

but for a macroscopic system and a small set F we may expect this limit to be approached only for enormous values of τ, while the phenomena of interest happen on a much shorter time scale. The spin echo counterexamplé warns us not to take this too much for granted.

A difficulty inherent in the choice of a finite τ is that Q now depends explicitly on the time parameter t of the state as well as on the state itself. This is due to the fact that the duration of the work cycles applicable to $\rho(t)$ is restricted to $[t,\tau]$. However, under some restrictive assumptions an approximate time homogeneity holds, as the following argument indicates.

Introduce one time scale (τ_1) for the work cycles and another (τ_2) for the study of the variation of $Q(\rho(t))$ with t. It is clear that the restriction $\tau_1 \gg \tau_2$ must be introduced in order that this variation shall not depend too strongly on the finite value of τ_1. $Q(\rho;\tau)$ must be a slowly decreasing function of τ in large

τ-intervals due to the monotonicity (6.10) and the existence of a lower bound. Assume that for a given small $\varepsilon > 0$ we can choose τ_1 and τ_2 such that

$$Q(\rho(s);\tau_1 - \tau_2) < Q(\rho(s);\tau_1) + \varepsilon$$

for all $0 \leq s \leq \tau_2$, i.e. there are no partial recurrences of period τ_1 to the accuracy given by ε. Then, for all $s \leq t \leq \tau_2$,

$$Q(\rho(t);\tau_1) - Q(\rho(s);\tau_1 - t + s)$$

$$\geq Q(\rho(t);\tau_1) - Q(\rho(s);\tau_1) - \varepsilon \geq - \varepsilon,$$

where the LHS is a time-homogenous quantity. In fact, the set of work cycles applicable to $\rho(t)$ ($D(\gamma) = [t,\tau_1]$) is the time translate of those applicable to $\rho(s)$ ($D(\gamma) = [s,\tau_1 - t + s]$). If the assumptions above hold while $Q(\rho(t);\tau_1)$ is still appreciably larger than the trivial limiting value (6.11) and ε is unobservably small, then the experimenter will see an effective irreversibility and time-homogeneity. Of course, in order that this property shall be interesting it should be possible to choose the same τ_1, τ_2 for a sufficiently large set of states.

The assumption of the existence of two different time scales is a common feature in discussions of irreversible statistical mechanics. Their interpretation is a bit different in the present context, due to the different starting point in the work cycles. The observation of a seemingly intrinsic irreversibility in S demands a clear separation of these time scales. On the time scale τ_1 no recurrence periods or effects of external perturbations should be seen. In order to be observed the relaxation processes, i.e. the increase of Q in time defined by (6.3), must take place in the time scale τ_2. The effective characterization of systems with the properties outlined above is of course an open problem.

In the following any restriction on the duration of the work cycles will often be implicitly included in the symbol F.

6.c Entropy functions

The quantity Q used above to describe the non-available energy and
the irreversibility depends on the origin of the work cycles in a
very inconvenient way. This deficiency is removed through the introduction of a set of entropy functions, which we will call col-
lectively "the thermodynamic entropy". The most complete definition , using also the reversible processes of Chapter 5, will only
be given in Chapter 10. Here the dynamics is still restricted to
the Hamiltonian type only.

In order to obtain a proper definition of the entropy functions a restriction must be made on the elements of the set F. The
simplest form of such a restriction was given in Chapter 3.b. A
less restrictive condition, but one which is more difficult to
check, is as follows.

Assumption: For every $H \in F$ the spectrum is discrete, there is a
non-degenerate ground level $E_0(H) = 0$ (as before). Furthermore

$$\{E(\beta,H); \beta \in (0,\infty)\} \supseteq \Delta(H) \tag{6.12}$$

$$\Delta(H) \equiv (0, \sup\{Q(\rho;F,H); \rho \in E(F)\})$$

From this follows that $\beta(u,H)$ introduced in Chapter 3.b is a well-
defined function of $u \in \Delta(H)$.

Using the condition of Chapter 3.b one can define for every
$\rho \in E_S$, $\gamma \in \Gamma(F)$

$$S(\rho;\gamma) = \int_0^E du\ \beta(u,H)$$

$$E \equiv \rho(H - \hat{W}(\gamma)), \qquad H \equiv 0(\gamma).$$

Alternatively, under the condition (6.12) we can define for every

$\rho \in E(F)$, $H \in F$

$$S(\rho;F,H) = \int_0^Q du\ \beta(u,H)$$

$$Q \equiv Q(\rho;F,H)$$

We can now give two different definitions of the thermodynamic entropy functions, the last being slightly more general.

Definition: The thermodynamic entropy defined by F is

$$S(\rho;F) = \inf\{\ S(\rho;\gamma);\ \gamma \in \Gamma(F)\} \tag{6.13}$$

$$= \inf\{\ S(\rho;F,H);\ H \in F\ \} \tag{6.14}$$

Proposition: The entropy functions have the following properties.

(1) $S_I(\rho) \leq S(\rho;F)$

for all ρ and every choice of F. Equality holds if $\rho \in G(F)$.

(2) If $\{\rho(t)\}$ is the time evolution of the state defined by any $\gamma \in \Gamma(F)$, then, for all $s \leq t$

$$S(\rho(s);F) \leq S(\rho(t);F).$$

(3) $S(\rho;F) \leq S(\rho;F')$

if $F' \subseteq F$. If F is so large that the work cycles generate all unitary transformations, then

$$S(\rho;F) = S_I(\rho),$$

which is the minimal entropy function.

(4) For any set $\{\rho_k \in E(F),\ \lambda_k > 0,\ \sum_k \lambda_k = 1\}$,

$$\sum_k \lambda_k S(\rho_k;F) \leq S(\sum_k \lambda_k \rho_k;F).$$

If equality holds, then $S(\rho_k;F)$ is independent of k.

Proof:

(1) $S_I(\rho) = S_I(T(\gamma)^*[\rho])$

due to the unitary invariance (3.1) of S_I, and

$\quad S_I(T(\gamma)^*[\rho]) \leq S(\rho;\gamma)$

due to the variational principle (3.6) and (2.6), which reads

$\quad \rho(T(\gamma)[H]) = \rho(H - \hat{W}(\gamma))$.

When using (6.12) and (6.14) this argument is restricted to optimal work cycles. When $\rho = \rho(\beta,H)$, $H \in F$, then, by the passivity property (6.6),

$\quad Q(\rho;F,H) = E(\beta,H)$,

and hence

$\quad S(\rho;F) \leq S(\rho;F,H) = S(\beta,H) = S_I(\rho)$.

(2) This follows from (6.3), (6.14) and the following relation which follows immediately from (3.7)

$\quad S(\beta',H) \leq S(\beta,H) \Leftrightarrow E(\beta',H) \leq E(\beta,H).$ \hfill (6.15)

(3) The first statement follows from (6.2), (6.14) and (6.15). If If F is so large that for every $\rho \in E_S$ there is $(\beta,H \in F)$ such that $\rho(\beta,H) \in \Omega(\rho,F)$, then

$\quad S(\rho;F) \leq S(\beta,H)$

follows from (2). Furthermore

$\quad S_I(\rho) = S(\beta,H)$

by the unitary invariance of S_I. From (1) follows that

$$S_I(\rho) = S(\rho;F).$$

(4) Put $\rho = \sum_k \lambda_k \rho_k$, $Q_k = Q(\rho_k;H)$, and leave out F in the notation. (6.5) says that

$$\sum_k \lambda_k Q_k \leq Q \equiv Q(\rho;H),$$

and (6.10) that for each k there is a unique $\beta_k = \beta(Q_k,H)$ with $Q_k = E(\beta_k,H)$. From (3.7) and (6.14) follows that

$$S(\rho_k) \leq S(\beta_k,H).$$

In the same way there is a unique $\beta = \beta(Q,H)$ such that

$$E(\beta,H) = Q, \qquad S(\beta;H) = S(\rho;H).$$

From (3.6) and the previous statements follows that

$$0 \leq \sum_k \lambda_k S_I(\rho(\beta_k,H)|\rho(\beta,H)) =$$

$$= S(\beta,H) - \sum_k \lambda_k S(\beta_k,H) + \beta(\sum_k \lambda_k Q_k - Q)$$

$$\leq S(\rho;H) - \sum_k \lambda_k S(\rho_k).$$

This holds for all $H \in F$. From (6.14) follows that

$$\sum_k \lambda_k S(\rho_k) \leq S(\rho).$$

Now let equality hold. By (6.14) there is for every $\varepsilon > 0$ a $H \in F$ such that

$$S(\rho;H) \leq S(\rho) + \varepsilon.$$

From the preceding arguments follows that

$$\sum_k \lambda_k S_I(\rho(\beta_k,H)|\rho(\beta;H)) \leq \varepsilon.$$

The LHS can be written as

$$\sum_k \lambda_k \int_{Q_k}^{Q} dy \int_y^Q dx \left[- \frac{\partial}{\partial x} \beta(x,H) \right].$$

For a finite system (without phase transitions!)

$$0 \leq - \frac{\partial}{\partial \beta} E(\beta,H) < \infty$$

for finite values of β. As all the β, β_k are finite we can find a constant C such that

$$- \frac{\partial \beta}{\partial u} \geq C > 0$$

for all $u \in [\min(Q_k,Q), \max(Q_k,Q)]$, and consequently

$$\sum_k \lambda_k (Q - Q_k)^2 \leq 2\varepsilon/C.$$

It follows that

$$S(\rho_k) \leq S(\rho;H) + \beta(Q_k - Q)$$
$$\leq S(\rho) + \varepsilon + \beta \left[\frac{2\varepsilon}{\lambda_k C} \right]^{\frac{1}{2}}$$

When $\varepsilon \to 0$ we obtain $S(\rho_k) \leq S(\rho)$. But then $\sum_k \lambda_k S(\rho_k) = S(\rho)$ implies that $S(\rho_k) = S(\rho)$ for all k.

The set of thermodynamic processes and entropy functions of this chapter are imperfect in some respects. For example, the optimal work cycles of different origin are not equivalent, and they do not in general define isentropic processes. The reason is that the reversible processes of Chapter 5 are not necessarily included in the dynamics used here. These defects will be repaired in Chapter 10, while the goal of this chapter was to show just how far one can proceed using the Hamiltonian dynamics of a finite system only.

In view of the influence of the recurrences on the concept of irreversibility and on the definition of the entropy functions, it is natural to ask what effect fluctuations in the system may have in this respect. As the work is defined in terms of ensemble averages, it is evident that fluctuations can not be used directly to produce work. Indeed, the fluctuations are relevant only when observations capable of resolving them are performed on the system, a case which is treated in Chapter 11. In the formalism given above it is only the recurrences predictable from the initial state which can be exploited to obtain work out of the system and which are a source of difficulty in defining the notion of irreversibility. It is perhaps useful to underline again that the entropy functions defined above are deterministic, non-fluctuating quantities.

We note that the idea of basing the definition of irreversibility on the notion of recoverable work is far from new. It was introduced by J. Meixner and collaborators in the context of linear irreversible thermodynamics [50]. In such models the linear response of the system to external forces is given by a phenomenological integral kernel which is designed to satisfy the passivity property. The work is defined from this response to be a quadratic form in the forces. Irreversibility can then be defined as non-recoverability of the work performed on the system to create the non-equilibrium state. The relation between energy and entropy is simple as the temperature is fixed. There is of course no microscopic dynamics in this case, the phenomenological dynamics being reversible or irreversible depending on the specification of the response kernel.

CHAPTER 7.   OPEN SYSTEMS

The problems and paradoxes which were met with in trying to define irreversibility for closed systems led many authors to the conclusion that open (non-Hamiltonian) systems provide better models for irreversible processes. This point of view was taken by Blatt [33], Mehra & Sudarshan [51] and Ingarden & Kossakowski [52] among others (see also Penrose [1], § 3.6). In this chapter the subject will be a thermodynamic description of open quantum systems in terms of entropy functions analogous to those defined in section 6.c. It turns out that for the simplest models of such systems, namely those given by Markovian master equations, such a description runs into unexpected difficulties.

7.a Markov description

Consider again a finite system S coupled weakly to a single heat bath $R = R(\beta)$. Chapters 4 and 5 dealt with the case where the Hamiltonian of S is constant in time or changes slowly compared to the relaxation induced by the S - R interaction. Now the fields F are allowed to act on S in the arbitrary way described in chapters 2 and 6. It will be shown how the thermodynamic entropy of a non-equilibrium state of S can be defined provided that the S - R correlations can be neglected. This condition is actually satisfied only in very idealized situations.

Instead of trying to solve the difficulties of defining the dynamics of the infinite system S + R and deriving the reduced dynamics of S (which is not feasible except in the WCL) the following assumptions are made. The set F denotes the Hamiltonians of S (as before), while the dynamics of R and the S - R interaction are the same for all values of the fields. Every $\gamma \in \Gamma(F)$ defines an evolution for $S + R$ and hence a non-Hamiltonian reduced dynamics for S as described in Appendix A.1. Under these circumstances the expression (2.7) for the work performed by $S + R$ is well-defined. Furthermore, the dynamics of $S + R$ is assumed to have the relaxation (return to equilibrium) property mentioned in Chapters 4 and 5.

Let F be fixed in this chapter and left out of the notation in most places. There is no need for a limitation on the duration of the work cycles as $S + R$ is taken to be infinite ($\tau = \infty$ in the notation of section 6.b). The reversible processes of Chapter 5 are included in the dynamics by the assumptions made above.

The objective of the open system approach is to treat the dynamics and thermodynamics of $S + R$ using only the reduced description given by the partial state ρ_S of S (defined in (2.7)) and the parameter β of R (which is fixed). To be more specific, the work functions (2.7) (and hence the derived thermodynamic entropy) are assumed to be functions of $\rho_S(t)$ only at each instant t. Note that (2.7) depends on the whole set of partial states $\{\rho_S(t); t \in D(\gamma)\}$, while we must demand that it depends on $\rho_S(t)$ only for all $\gamma \in \Gamma(F)$ with $D(\gamma) \subseteq [t, \infty)$. This hypothesis is clearly satisfied if the reduced dynamics of S has the *Markov property* that $\rho_S(t)$ determines $\rho_S(u)$ uniquely for all $u > t$ and all $\gamma \in \Gamma(F)$ (see Appendix A.2 for a more complete definition). The Markov assumption is also necessary if the reduced description is non-redundant in the sense that the work functions (2.7) for all $\gamma \in \Gamma(F)$ with $D(\gamma) \subseteq [t, \infty)$ determine $\rho_S(t)$ uniquely. This is true if F is large enough relative to S, e.g. if $F = 0_S$, which may be a natural assumption if S is a small system, say of atomic dimensions. If S is a system of macroscopic size, then the work functions can not be expected to define ρ_S. However, even in this case the Markov assumption is the simplest, indeed the only practical way to ensure that the work functions have the demanded property. Consequently it will be taken for granted in most of this chapter and in several places further on.

The Markov property of the reduced dynamics implies that in a complete description of $S + R$ the $S - R$ correlations are inconsequential for the work functions. Furthermore, the $S - R$ interaction and the consequent energy transfer must leave the equilibrium state of the infinite system R unchanged. This means that for the prediction of the future state of S and the work functions, there is the following observational equivalence for the states of $S + R$:

$$\rho_{S+R} \simeq \rho_S \otimes \rho_R(\beta) \tag{7.1}$$

where ρ_S is defined in (2.7) and $\rho_R(\beta)$ is the Gibbs state of R. In the equilibrium states of $S + R$ the correlations can be neglected if the $S - R$ interaction is sufficiently weak:

$$\rho_{S+R}(\beta,H) \simeq \rho_S(\beta,H) \otimes \rho_R(\beta) \tag{7.2}$$

The assumptions introduced above may be formulated in the following way: It is possible to choose the $S - R$ boundary in such a way that the $S - R$ correlations can be neglected. That this idealization leads to trouble will be seen below.

The assumption that the two sides of (7.1) are equivalent for the prediction of the evolution of S gives a basic inequality for the I-entropy. Let the time evolution give

$$T^*(t - s)[\rho_S(s) \otimes \rho_R(\beta)] = \rho_{S+R}(t) \simeq \rho_S(t) \otimes \rho_R(\beta)$$

The increase in the I-entropy of $S + R$ given by (3.10), which is due to the neglect of the $S - R$ correlations, and the change in the I-entropy of R given by (5.1) give, for all $s < t$ and all $\gamma \in \Gamma(F)$ with $D(\gamma) = [s,t]$, $O(\gamma) = H$,

$$\Delta S_I^{S+R} = \Delta S_I^S + \beta \Delta E^R = S_I(\rho(t)) - S_I(\rho(s)) -$$

$$- \beta\{W(\gamma,\rho(s)) + (\rho(t) - \rho(s))[H]\} \geq 0 \qquad (7.3)$$

where ρ stands for ρ_S from now on.

7.b Available work and entropy

The energy Q defined in Chapter 6 now refers to $S + R$ and is infinite. It is convenient to use instead the finite quantity (6.7)

$$A(\rho_{S+R};H) = \sup\{W(\gamma,\rho_{S+R}); \gamma \in \Gamma(F), O(\gamma) = H\}.$$

Note that this work is extracted from $S + R$. From (3.8) follows

$$0 \leq \beta A(\rho_{S+R};H) \leq S_I(\rho_{S+R}|\rho_{S+R}(\beta,H))$$

where the RHS can be written, due to (7.2),

$$S_I(\rho_{S+R}|\rho_{S+R}(\beta,H)) = S_I(\rho|\rho(\beta,H)) + S_I(\rho_R|\rho_R(\beta)).$$

The part belonging to R can not be used to produce work, by the assumptions above. In fact, as R is assumed not to deviate significantly from equilibrium, (5.2) gives

$$S_I(\rho_R|\rho_R(\beta)) = 0$$

in the thermodynamic limit. Hence, expressed in the partial state of S

$$0 \leq \beta A(\rho;H) \leq S_I(\rho|\rho(\beta,H)). \qquad (7.4)$$

Now let ρ be one of the Gibbs states of S. Due to the assumed relaxing properties, the reversible processes of Chapter 5 will provide a *Carnot cycle* as an optimal work cycle. In fact, if $\rho = \rho(\beta,H_1)$, $H_1 \in F$, then equality holds in (7.4):

$$\beta A(\rho(\beta,H_1);H) = S_I(\rho(\beta,H_1)|\rho(\beta,H)) \qquad (7.5)$$

due to the existence of the optimal work cycle of origin H which is the "limit" as $\tau \to \infty$ of the set $\{\gamma_\tau\}$,

$$\gamma_\tau: H(t) = H \text{ for } t = 0,\tau$$

$$= H_1 + \frac{t}{\tau}(H - H_1) \text{ for } t \in (0,\tau),$$

(7.6)

and which gives equality in (7.3) by the argument of Chapter 5 and hence in the second inequality of (7.4).

We can now define the *thermodynamic entropy* of the open system S in the following way (F and β are fixed)

Definition:

$$S(\rho) = S(\beta,H) + \beta[\rho(H) - E(\beta,H) - A(\rho;H)]$$

(7.7)

Proposition: $S(\rho)$ has the following properties.

(1) $S(\rho)$ does not depend on the origin H of the cycles.

(2) $S_I(\rho) \leq S(\rho)$ for all ρ. Equality holds when $\rho \in G(\beta,F)$.

(3) For any $\gamma \in \Gamma(F,H)$, $D(\gamma) = [s,t]$,

$$S(\rho(s)) \leq S(\rho(t)) + \Delta S^R$$

(7.8)

$$\Delta S^R = -\beta\{W(\gamma,\rho(s)) + (\rho(t) - \rho(s))[H]\}.$$

The isentropic processes, which are those where equality holds in (7.8), correspond to optimal work cycles.

(4) For all sets $\{\rho_k \in E_S, \lambda_k \geq 0; \sum_k \lambda_k = 1\}$,

$$\sum_k \lambda_k S(\rho_k) \leq S(\sum_k \lambda_k \rho_k).$$

Proof: (1) By the assumed relaxation property there are optimal work cycles $\gamma \in \Gamma(F,H)$, $\gamma_1 \in \Gamma(F,H_1)$, such that

$$T(\gamma)^*[\rho] = \rho(\beta,H)$$

63

$$T(\gamma_1)^*[\rho] = \rho(\beta, H_1).$$

(The dynamical maps are now non-unitary!) Define a work cycle $\gamma' = \gamma_1' * \gamma_2$ where $\gamma_1' \simeq \gamma_1$, $O(\gamma_1') = H$ and γ_2 is a reversible Carnot cycle of the form (7.6). From (7.5) follows that

$$W(\gamma',\rho) = W(\gamma_1',\rho) + W(\gamma_2,\rho(\beta,H_1))$$

$$= W(\gamma_1,\rho) + (\rho - \rho(\beta,H_1))[H - H_1] +$$

$$+ \beta^{-1} S_I(\rho(\beta,H_1) | \rho(\beta,H)) \ ,$$

and hence that

$$\beta A(\rho;H) \geq S(\beta,H) - S(\beta;H_1) +$$

$$+ \beta\{A(\rho;H_1) + \rho(H - H_1) + E(\beta,H_1) - E(\beta,H)\}$$

Exchanging H and H_1 gives the opposite inequality, hence equality holds.

(2) Using the relaxing property as in (1), we can put

$$\lim_{t \to \infty} \rho(t) = \rho(\beta;H)$$

and $\rho(s) = \rho$ in (7.3) to obtain

$$S_I(\rho) \leq S(\beta,H) + \beta\{\rho(H) - E(\beta,H) - W(\gamma,\rho)\}$$

for all $\gamma \in \Gamma(F,H)$ and hence the inequality. If $\rho = \rho(\beta,H)$, $H \in F$, then there is, due to the passivity of ρ, the trivial optimal work cycle $H(t) = H$ for all t, hence equality holds.

(3) (6.3) is equivalent to

$$A(\rho(t);H) + W(\gamma,\rho(s)) \leq A(\rho(s);H)$$

for s < t, which, with (7.7), gives (7.8) directly. Equality in (7.8) corresponds to equality in (6.3), i.e. to optimal cycles.

(4) This follows directly from (6.5), (6.7) and (7.7).

Note that the definition (7.7) can be written in the following way, introducing a relative entropy for S (compare (3.9) and (7.5))

$$S(\rho|\rho(\beta,H)) = \beta A(\rho;H).$$

The inequality (7.8) is the second law for $S + R$ expressed through the reduced description of S. Obviously it does not make sense unless the dynamics of S is Markovian. We now turn to the problem of finding a justification for this assumption.

7.c Master equation models

By assumption, S stays near equilibrium when the fields vary slowly compared to the rate of relaxation. It is then possible to find the deviation from reversibility using the WCL Markovian master equation. This was done by Spohn [53] and Spohn & Lebowitz [24], who defined a positive entropy production and discussed its properties. Alicki applied this concept to models of heat engines and heat conduction [54,55].

For any $\gamma = \{H(t)\} \in \Gamma(F,H)$ with $D(\gamma) = [0,\tau]$, which defines the evolution $\rho(t)$ of the state, we write

$$\Delta S(\gamma,\rho(0)) = \int_0^\tau dt \, \sigma(t)$$

$$\sigma(t) = \frac{d}{dt}[S_I(\rho(t))] + \frac{d\rho}{dt}(t)[\ln\rho(\beta,H(t))] \qquad (7.9)$$

A simple calculation gives

$$\Delta S = - \beta W(\gamma,\rho(0)) + S_I(\rho(0)|\rho(\beta,H)) - S_I(\rho(\tau)|\rho(\beta,H)).$$

In the WCL the state of S will relax to equilibrium, i.e. by choosing τ large enough we have

$$\rho(\tau) \simeq \rho(\beta,H)$$

$$\Delta S \simeq - \beta W(\gamma, \rho(0)) + S_I(\rho(0)|\rho(\beta,H)).$$

When $\rho(0) = \rho(\beta,H_1)$, $H_1 \in F$, then it follows from (7.3) and (7.5) that ΔS is the total I-entropy increase

$$\Delta S = \beta[A(\rho(0);H) - W(\gamma,\rho(0))] = \Delta S_I^{S+R} \geq 0$$

If $H(t)$ varies slowly compared to the rate of relaxation, then the WCL gives a master equation (A.12), where the dissipative part L_d of the generator is now a function of t [54]. For each fixed t the generator is of the form given in (A.7) and has the property (A.8) of annihilating the Gibbs state corresponding to the instantaneous value of the Hamiltonian, i.e. for all t

$$L(t)^*[\rho(\beta,H(t))] = 0, \tag{7.10}$$

Then the expression (7.9) can be written

$$\sigma(t) = - \frac{d}{ds} S_I (e^{sL^*(t)} [\rho(t)] | e^{sL^*(t)} [\rho(\beta,H(t))]) \big|_{s = 0}$$

From the H-theorem (3.5) and (A.13) it follows that for all t

$$\sigma(t) \geq 0 \tag{7.11}$$

σ is the I-entropy production defined by Spohn [53]. Equation (7.11) means that the I-entropy is a consistent measure of the irreversibility in the reduced dynamics of S given by the WCL master equation as long as (7.10) holds.

When the fields are allowed to vary at a rate which is not slow compared to the relaxation, then σ can not be expected to satisfy (7.11). Provided that the variation takes place on the rescaled time of the WCL the results of Davies and Spohn [41] still give a semigroup evolution. Now the form of the generator (equation (6) in [41]) is more complex than in the previous case and it does not satisfy (7.10) in general. Consequently the H-theorem (A.13) can no longer be interpreted in terms of the I-entropy

66

production σ.

For arbitrary variation rates there are no rigorous results as the rescaling of the time in the WCL does not permit a treatment of this case. The evolution of S will not be given by a semigroup in general, i.e. there will not be a family of dynamical maps satisfying (A.11). This means that the dynamics can not have the Markov property of section 7.a and Appendix A.

Instead of trying to derive an approximate Markovian reduced dynamics for S, we may attempt to find model dynamics with the Markov property, chosen only for mathematical simplicity and tractability. The question then arises if the model is consistent with thermodynamics. A minimal condition is the passivity of the stationary "equilibrium" states obtained as limits when $t \to \infty$ and the fields have constant values. If the passivity property is not fulfilled and if there is a finite relaxation rate to the stationary state, then an infinite amount of work can be extracted from the open system. This work comes from the heat bath, and it is evident that no thermodynamic entropy can be defined in this case.

For the models of irreversible quantum dynamics called *quantum stochastic processes* [56] the Markov property is given in terms of the *quantum regression theorem* (A.9). This describes the higher order correlations for repeated measurements as well as the response to the time-dependent fields. This property implies that the dynamical maps always form a semigroup (see (A.11)). The Markov property is equivalent to a monotonic loss of information contained in the state of the system which is expressed in the H-theorem (A.13) [25].

The evolution of such Markovian systems is given by a master equation (A.12) where the dissipative part L_d of the generator is now *fixed, independent of* H(t). Then the stationary state for a given constant H is not the Gibbs state $\rho(\beta,H)$ in general (except for $\beta = 0$). This in itself is no disaster, but together with the evolution (A.12) for a time-dependent Hamiltonian, it implies that the system is not passive when $\beta > 0$. This is easy to see in the example of the Bloch equations in NMR when the magnetic field can be chosen arbitrarily strong. The details of the calculations are given in Appendix A.3. It is found that in this particular example the rate at which the system can perform work is of the same order of magnitude as the rate of dissipation. This result is clearly in conflict with the passivity property of the state of the heat bath. It is of course due to the fact that the derivation of the master equation (A.12) is not valid for the relevant work cycles. It is evident that if only small deviations from a particular Gibbs state (defined by a fixed H_0) are considered, then this non-passivity will be a small second order effect. These Markovian models also have other properties inconsistent with the equilibrium state of the reservoir, and these are described in Appendix A.

A model which is consistent with thermodynamics is obtained if the WCL form of the dissipative part of the generator satisfying (7.10) holds, by definition, for arbitrary $H(t)$. This type includes the modified Bloch equations of NMR [58]. It is quite similar to the classical probabilistic model discussed in [59]. The validity of (7.11) implies that the model is passive. If $F = 0_S$, then the thermodynamic entropy for S is the I-entropy

$$S(\rho,F) = S_I(\rho),$$

as every state is a Gibbs state.

This model, however, has another disease. The instantaneous adaption of the dissipative part of the generator to the value of the Hamiltonian involves an implicit assumption of an infinitely fast relaxation in the $S - R$ correlations. This assumed behaviour expresses itself in a highly singular dependence of σ on the state of S , making the entropy production infinite for most initial states [53]. Furthermore, the dynamics is so singular that it is not possible in general to specify in a consistent way how S will interact with other systems, in particular with any measuring apparatus. One may express this property as a lack of dynamical consistency, which can be seen in the following way. Choose H_1, $H_2 \in F$ and let

$$T_i(t) = \exp[tL(H_i)], \quad i = 1,2,$$

be the semigroups of dynamical maps defined by H_1 and H_2, respectively, through (A.5) where the generator $L = L(H)$ for each value of H has the WCL form satisfying (A.8). Then, by the Lie - Trotter formula [60]

$$T(t) = \lim_{n \to \infty} [T_1(t/2n) \cdot T_2(t/2n)]^n$$

is the semigroup generated by

$$L = \frac{1}{2}(L(H_1) + L(H_2)).$$

On the other hand, the Lie - Trotter formula applied to a sufficiently regular Hamiltonian evolution of $S + R$ would give that $T(t)$ represents the evolution generated by

$$L = L(H), \qquad H = \frac{1}{2}(H_1 + H_2).$$

As it follows from the explicit formula for L [15,24,44] that

$$L(H) \neq \frac{1}{2}(L(H_1) + L(H_2))$$

in general, there is a contradiction which shows that the dynamics of $S + R$ must be very singular in this model. This singularity means that the higher order correlations for the reduced dynamics of S which are defined in [56] can not be defined. Consequently there is no proper stochastic process associated with this model.

The inconsistencies of the two models descibed above pose no problems in the WCL, where the rate of variation of H(t) must be taken to be on the scale of the rate of dissipation. The Markovian master equation model is also consistent with thermodynamics as long as the energy transfer from R is insignificant or $\beta = 0$. In the general case, however, there seems to be no Markovian description of the relaxation to equilibrium of open quantum systems which is consistent with the passivity of the equilibrium states.

The only way of avoiding these problems seems to be to accept a non-Markovian evolution for the open system S. The "memory" of the dynamics involves the information that $S + R$ was in an equilibrium state at a specified moment in the past, complete with the $S - R$ correlations defined by the interaction Hamiltonian. The evolution of S in a time-dependent field has to be consistent with the passivity of this equilibrium state. This condition implies that the dynamics can have no strict Markov property except when the correlations and hence the interaction and the dissipation are absent. It seems that even the stationarity property postulated for the more general non-Markovian quantum stochastic processes introduced in [56] can not be fulfilled. One can see this as a feature of the *holistic property* of quantum dynamics: There is no consistent description of the state of a quantum system in terms of the partial states of interacting subsystems.

For classical systems and commutative stochastic processes the problems described above do not occur. A stochastic process does not define the response of the system to external forces in the commutative case, but it can describe the spontaneous thermal fluctuations. In linear response theory the fluctuation-dissipation theorem couples the two aspects in a way which is consistent with thermodynamics [50,61], and it is not difficult to see that in particular it imposes the passivity property. Of course, such models always describe small deviations from a given equilibrium state, a case which did not lead to problems in the quantum case either.

70

For more general models, especially the present ones which are based on the microscopic dynamics, it seems necessary to distinguish between the following two types of processes:

(1) The transient relaxation to equilibrium conditioned by the preparation of a state far from equilibrium in a large but finite (sub-) system. This case demands a thermodynamic description, but there is no obvious stationarity property here.

(2) The small thermal fluctuations in an open system, which can often be described by a stationary stochastic process, e.g. as defined by a Markovian master equation.

The apparent contradiction between randomness and thermodynamics for quantum systems is inconvenient as it restricts the set of simple and selfconsistent models. However, in physical applications there is a natural exclusiveness of the two points of view. The modelling of the dynamics by a stochastic process is relevant for systems with a few degrees of freedom. For them the physical entropy is insignificant due to the small value of Boltzmann's constant. For systems of macroscopic size the assumption of a Markov property or a more general stochastic model does not really correspond to something observable, as the information needed to specify the microscopic state of the system is much too large.

CHAPTER 8. EXTERNAL PERTURBATIONS

Even when the system S is separated from the heat baths, it will
be subject to perturbations from that part of the external world
(called X) over which the experimenter has no effective control.
We already noted that when S is in a Gibbs state corresponding to
the actual value of the Hamiltonian, then the state is stable
under small random perturbations of the dynamics. This stability
can be expressed by the bound (3.15) on the I-entropy production
provided by the energy transfer. Non-equilibrium states do not
have this stability property and there is every reason to believe
that even very small perturbations may be amplified by the intrinsic dynamics of S and have a sizeable influence on the evolution
of the states. This influence will be seen as an increase in the
I-entropy, and, in the formalism of Chapters 2 and 6, as a destruc-
tion of any group property of the mobility semigroup. The object
of the present chapter is to give some justification for these
statements. However, a rigorous mathematical treatment of this set
of problems is lacking for quantum systems and even for classical
Hamiltonian systems, while a bit more is known about some other
classes of classical dynamical systems.

8.a Models of the perturbations

The perturbations include fluctuations in the fields F which define
the set F. If F is treated correctly as a quantum field, then there
will be, at least, zero point fluctuations. If F serves to couple
different unperturbed energy levels of S , which it must do if the
work cycles in $\Gamma(F)$ shall be able to perform work on S, then there
will be an interaction between S and F which provides a line width
for the energy levels. A similar conclusion holds if F represents

a container for S made up of atomic constituents. We conclude that there is in any experimental setup defining S and F a minimal interaction with the external world which perturbs S.

In contradistinction to the interaction with the heat baths studied in Chapter 7, the relevant case here is that the interaction is so weak that $S + X$ will not reach equilibrium during the experiment. In fact, the total exchange of energy can be neglected as the experimenter would reject the data if he found an energy change in S unaccounted for by the reproducible interactions with the heat baths and the fields. In addition we have to assume that the energy exchange with each part of X with a definite temperature is practically zero. Otherwise S could be driven by the external world in an uncontrolled manner while the average energy of S stays constant.

The first problem is to find a mathematical model for the action of the external perturbations. We assume that $S + X$ will not reach equilibrium during the preparation of the initial state of S. Furthermore, the macroscopic state of X and the precise form of the $S - X$ interaction will in general be unknown. Consequently the $S - X$ correlations are not reproducible. This does not mean that the correlations can be neglected completely. Sometimes a part of the perturbing influence of X can be removed by suitable manipulations of the fields, in spite of this lack of information. In NMR experiments *averaging* (multiple pulse) techniques will do just that for certain types of perturbations, e.g. not too fast local fluctuations in the applied magnetic field [62]. It was pointed out in [63] that this effect is typical for non-Markovian evolutions. Consequently, a Markov description (in the sense of Appendix A and section 7.a) of the evolution of the perturbed system S, with the attendant neglect of the $S - X$ correlations, must not be taken for granted. In spite of this caveat, it seems reasonable to take a Markovian evolution as a model of the "irreducible part" of the effect of the perturbations, i.e. that part which can not be removed by manipulations of the external fields.

As an example of such Markovian models we have mentioned the Hamiltonians with white noise perturbations treated in [30,31] and which lead to semigroup evolutions where the dissipative part of the generator has the form (3.11). For these models the I-entropy is non-decreasing (equation (3.12)). They represent the action of singular reservoirs of infinite temperature, as there is no quantum white noise for finite temperatures (see [64-66] and Appendix A). This means that the validity of such a model is restricted to time intervals so short that the energy exchange between S and X is insignificant. By (3.15) the perturbation of the equilibrium states can then be neglected. For this type of model (with $\beta = 0$!) there is no problem with the incompatibility of Markov models (for $\beta > 0$) and thermodynamics described in Chapter 7. This is true even if a finite energy exchange is taken into account, a fact which seems to be a good reason to make this choice for a model of the perturbations. For classical systems the corresponding model is a diffusion term added to the deterministic evolution in phase space.

8.b Classical systems

Provided with a stochastic model of the external perturbations one may try to find an answer to the following problem: What are the consequences of these perturbations (and the consequent increase in S_I) for the irreversibility of the dynamics of S?

The Poincaré recurrence theorem led some investigators to the conclusion that the uncontrollable perturbations due to the external world is essential for irreversible behaviour even for systems which are essentially closed (in the sense of having a negligible energy exchange with the environment). This line of ideas goes back at least as far as E. Borel [67,68]. He calculated the effect

of small fluctuations in the gravitational field (due to events
on distant stars) on the trajectories of molecules in the hard
sphere gas model. He found that the collisions of the molecules
lead to an *exponential divergence* of initially close trajectories
and argued that they rapidly become essentially unpredictable as
a result of the perturbations. Similar arguments have been presented numerous times (see e.g. the review by Berry [69]). This
exponential *instability* of the dynamics was an important feature
of the work of Krylov [34] and it is the basis for the work of
Sinai on billiards and the hard sphere gas [49,70].

It has been claimed, e.g. in connection with the description
of turbulence, that external perturbations are not necessary in
order to have a stochastic motion in classical dynamical systems
[71]. Instead an instability of the motion can give rise to a sensitive dependence on initial conditions which suffices to make the
motion unpredictable [71,72]. The term *chaos* is often used to describe motion with unpredictable (stochastic) properties though
the equations of motion are deterministic.

The long-term unpredictability of the motion can be taken to
mean that in an infinite sequence of repeated observations on the
system, each additional observation will give a well-defined quantity of new information. The resulting creation of information ad
infinitum is possible due to the following mathematical fact. For
a classical dynamical system an initial state which is given by a
measure on phase space absolutely continuous relative to Lebesgue
measure contains an infinite amount of uncertainty, i.e. to specify one point in the phase space takes an infinite amount of information. The instability of the dynamics can amplify this uncertainty to a "macroscopic" or "observable" level at an asymptotically constant rate [72].

In ergodic theory the asymptotic information gain is given

by the Kolmogorov - Sinai entropy invariant [49,73]. Systems with
the K-*mixing* property have a positive value for the KS-entropy
and consequently a genuinely unpredictable quality in spite of
the formal determinism of the dynamics [73].

The instability of the dynamics can alternatively be formulated in terms of the *Liapounov exponents* which give the asymptotic exponential divergence of the trajectories mentioned above
[71,74]. Systems with this exponential instability are often
called *hyperbolic*. For smooth dynamical systems with smooth invariant measures there is a relation between the Liapounov exponents and the KS entropy given by a simple formula due to Pesin
[71,74].

It can not be expected that realistic models are uniformly
chaotic over the whole phase space. Numerical work on simple systems indicates that the phase space will be divided into subregions
where the KS entropy (and the Liapounov exponents) take on different values. For Hamiltonian systems there seems to exist a threshold in the energy where non-zero values, as calculated by numerical approximation, first appear. This threshold is often called
the *stochastic transition* [46].

In the case of finite quantum Hamiltonian systems with dis-
drete spectra there can be no asymptotic characterization of instability properties on the lines described above. No non-trivial
generalizations have been given of the concepts of KS entropy or
K-mixing for such systems, although several suggestions for a
quantum KS entropy have been made for infinite or open systems
[56,75-77]. Trajectories are not defined for quantum systems and
neither are the Liapounov exponents. Another way of seeing the
lack of randomness is by noting that for any realistic quantum
system a state of finite energy has a finite I-entropy and
thus it can not provide the infinite uncertainty needed for

a genuinely non-deterministic stochastic process in $t \in (0,\infty)$. In fact, we already noted that the motion is almost periodic and at best ergodic. This is also true for a finite completely integrable classical Hamiltonian system where the motion is quasi-periodic (i.e. almost periodic with a finite number of frequencies, also called conditionally periodic)[69,73]. For these systems there is, at least in a formal sense, a good long-term predictability of the motion.

When searching for some property of the chaotic classical systems which it might be possible to generalize to the quantum case, their sensitive dependence on external perturbations (in the spirit of Borel) seems to be a likely candidate. From simple models one can see a direct relation between the rate of exponential divergence of trajectories and the increase in the I-entropy which results from the addition of a small diffusion term to the dynamics (Appendix B). In some approximation the I-entropy will show a linear increase in time with a coefficient which is intrinsic to the system, i.e. it is independent of the strenth of the noise term. This approximatively linear increase can obviously only hold during a finite time interval, until the system approaches a stationary "equilibrium" distribution. For dynamical systems with a sufficiently strong hyperbolic property the stationary states will have a stability property under small random perturbations of the dynamics [78,79]. This stability is in some ways reminiscent of that of the Gibbs states described in Chapter 3.

The example in Appendix B indicates that the value of $S_I(\rho(t))$ will depend on the strength ε of the perturbation and the time parameter t in a characteristic way. This implies that the system can be considered as closed, in the sense of having a constant I-entropy to a given accuracy, during an interval $(0,t)$ if the noise satisfies a bound (valid for all t) of the form

$$\varepsilon \leq C_1(\rho) \exp[- h(\rho)t - k(\rho,t)] \qquad (8.1)$$

where $\rho = \rho(0)$, $h(\rho) > 0$ and $k(\rho,t)$ is a non-decreasing function of t which has a finite limit as $t \to \infty$ (this is (B.7)). The form of (8.1) clearly underlines the practical difficulties of protecting a system with an exponential instability from external perturbations.

For classical dynamical systems with quasi-periodic motion the effect of perturbations is more complex. We already noted that the difference between chaotic and quasi-periodic systems is well-defined and absolute due to the different asymptotic behaviour. However, from the physical and computational points of view the distinction becomes blurred due to noise and truncation errors, respectively. In fact, a stochastic trajectory may be very difficult to distinguish in practice from a regular one of an extremely long period.

Classical Hamiltonian systems may have a local, approximately exponential divergence of trajectories [80]. This local instabilty will not necessarily lead to a global instability. The local exponents may be averaged out to give the asymptotic linear divergence of trajectories which is characteristic of regular (non-stochastic) motion [81,82]. This averaging can be expected to take place over a time interval of length at least comparable to the recurrence periods. However, the details of this process does not seem to be understood at all. Some computer experiments have been performed which may serve as a guide for the imagination [83].

Let a quasi-periodic system with a local instability be in a state far from equilibrium. When the dynamics is perturbed be external noise, the I-entropy can be expected to increase approximately linearly with time in a certain interval of length τ, say, for which the exponential divergence of trajectories is relevant. From the example in Appendix B it can be conjectured that the

I-entropy production is some average of the local exponents similar to Pesin's formula [74]. The condition for having a closed system of approximately constant I-entropy is then still of the general form (8.1). The quasiperiodicity can not be seen unless the noise satisfies (8.1) with t = τ. Provided that the local exponents are not averaged out in a much shorter time, long recurrence periods will give a very low bound indeed.

Over a time span longer than the recurrence periods, the regular nature of the system will become apparent. A logarithmic increase of the I-entropy is then expected from the linear divergence of trajectories, again provided that the system is still far from equilibrium. This is the behaviour in the quasiperiodic regime indicated by the model in Appendix B (see (B.9)). The bound on ε will then be of the form (B.10)

$$\varepsilon^2 \leq C_2(\rho) \ t^{-1} \qquad\qquad (8.2)$$

where C_2 is independent of t. C_2 can be expected to be exponentially small when the local exponents and the recurrence periods are large.

8.c Quantum systems

We now return to the case of finite quantum systems. Here little of relevance is known avout the possible types of behaviour under small random perturbations. Thus the picture which can be sketched for this case must unavoidably be even more conjectural than that given above.

There seems to be no a priori reason to believe that quantum systems with discrete spectra differ in principle from completely integrable classical systems in their sensitivity to perturbations. In both quantum and classical cases almost periodic evolutions with very long recurrence periods may exist, which do not differ from the unstable dynamics described above in their observable behaviour on a realistic time scale. The properties which could characterize such quantum dynamics has been discussed in the extensive recent work on the very vaguely defined concept of *quantum chaos* (or quantum stochasticity) [80,84,85]. This concept is a label for a large number of different and not obviously compatible attempts to find a quantum analogue of the classical exponential instability. The theoretical work has largely been impelled by the experimental observation of seemingly chaotic phenomena in quite small molecules [84].

The most obvious condition for the observation of a quantum instability, in a system with a given Hamiltonian H, is that the relevant time scale is shorter than the periods of spontaneous recurrences. The periods can increase extremely fast with the size of the system (see Chapter 9 for an estimate). Consequently this condition is a very weak one for large systems. Note that, due to the rescaling of the time parameter, the WCL treatment of the present problem by its very nature deals with a time scale long compared to the recurrence periods. This is not the interesting case here.

A considerably stronger condition for quantum chaos is obtained through the following argument (for which no rigour is claimed). We know that the K-mixing property implies a Lebesgue spectrum for the correlation functions of a classical system [73]. In the case of a quantum system the spectrum of the correlation function

$$r(t) = \rho(T(t)[A]^+ A)$$

(see (A.16)) is given by the resonance frequencies

$$\{\omega_{kl} = \hbar^{-1}(E_k - E_l); \ E_k \in \text{Sp } H\}.$$

Hence, a quantum system can only be expected to mimic the classical

unstable motion during a time too short to reveal the discreteness
of this spectrum. There is then an uncertainty relation between
the time scale τ and the maximal separation δE between neighbour-
ing energy levels

$$\tau \cdot \delta E < \hbar . \tag{8.3}$$

If an average or minimal level separation is chosen for δE we may
expect to see a combination of regular and chaotic behaviour.

On the other hand, in order to see any evolution in S at all,
the total energy uncertainty in the given state must satisfy the
opposite inequality. The observation of a chaotic evolution certainly demands that the observation time is several times the
"relaxation time" of the correlation function given by the width
of the energy distribution, i.e.

$$\tau [\rho [(H - \rho(H))^2]]^{\frac{1}{2}} \gg \hbar .$$

In order to show a stochastic behaviour over a considerable time
interval, quantum systems should thus have a large number of (occupied) energy levels. The levels should have an even distribution
on the energy scale, with a fairly uniform spacing between nearest
neighbours. Stochastic behaviour is believed to be associated with
"level repulsion" leading to non-degenerate levels and avoided
crossings when the fields are varied [84,85]. As an antithesis of
this behaviour, quantum systems corresponding to regular classical
systems may have highly degenerate spectra. There are then some
large level separations which can lead to violations of (8.3) and
even to recurrences of short periods. This degeneracy or almost
degeneracy of the energy levels can be coupled to the existence of
conserved or nearly conserved observables besides the Hamiltonian.
Recall that for classical systems the existence of a full set of
analytic constants of motion (complete integrability) is equivalent
with regularity (quasi-periodic motion) [69].

The question is now: How kan a quantum stochastic property be defined in terms of the I-entropy production induced by the external perturbations? First note that the approximate energy conservation in S demands not only that the perturbation parameter ε of the generator in (3.11) is small, but also that the coupling of the unperturbed energy levels provided by the operators $\{V_i\}$ does not lead to observable "quantum jumps" in S. Thus, it must be assumed that the generator can only couple levels inside energy intervals of a given length, which may be at most of the order of the total allowed energy transfer. Also note that the inequality (8.3) implies that the coupling of nearby energy levels poses no problem as regards energy conservation.

The estimate (3.16) supports the idea that systems with very dense energy levels can be highly sensitive to perturbations, in the sense that ΔS_I is not effectively bounded by the energy transfer. However, if we estimate the energy transfer by an expression of the form

$$\Delta E \simeq \mathrm{const} \cdot \varepsilon^2 t,$$

we see that the condition on ε for ΔS_I to be small is of the form (8.2) rather than (8.1). The reason is that the Hamiltonian dynamics of S and the non-diagonal terms of the state (in the eigenbasis of the Hamiltonian) have not been taken into account in (3.16). This must be done in order to obtain a bound of the form (8.1), which is characteristic of the exponential instability for classical systems. Let us write (3.16) in the form

$$\Delta S_I(t) \simeq \mathrm{const} \cdot \ln N(t) \tag{8.4}$$

where $N(t)$ represents the number of (pure) states over which the perturbations smear an initially pure state during $(0,t)$. Then the bound (8.1) is relevant if $N(t)$ is of the form

$$N(t) \simeq \text{const} \cdot \varepsilon^2 \exp(ht), \qquad\qquad (8.5)$$

where h is a constant. This form corresponds to a linear increase in the I-entropy with time. An exponential increase in the number of states smeared by the perturbation can thus be seen as a quantum version of the classical exponential instability. I can not claim that a complete justification of (8.4) and (8.5) is possible at present, in the sense of giving sufficient conditions for them to hold. Only three necessary prerequisites for this behaviour to exist will be described (apart from a restriction of the form (8.3)), as well as the consequences of their non-fulfillment.

(1) The dissipative generator modelling the noise must not commute with the Hamiltonian H which describes the intrinsic dynamics of S. If they do commute, then the entropy production is clearly the same as if we had put H = 0. This commutativity holds for the WCL form of the generator (Appendix A). Obviously this case can not describe an instability property intrinsic to S. Instead we expect an increase in time of the I-entropy which is approximately logarithmic, corresponding to a linear increase in N(t), just as in the regular classical case (compare (B.9)). The bound (8.2) is then the relevant one.

(2) In order to have a large I-entropy production, the perturbation must couple many, preferably all, of the energy levels which are allowed by the approximate energy conservation. This means that they should not satisfy any selection rules associated with symmetries of H. Instead, even small perturbations should suffice to mix nearby levels. This picture seems to tally well with the notion of "irregular spectra" discussed by Percival [86]. An example of a system with selection rules is the following. Let S consist of several non-interacting subsystems and assume that the perturbations do not couple them either (they are "localized"). The spectrum of S may be dense "by accident" but the perturbations will only couple levels belonging to a single subsystem with a more sparse spectrum.

If the subsystems have regular behaviour, then we can not expect
to see chaos in S either.

(3) The total increase in the I-entropy is bounded by the entropy
of the equilibrium state of the given energy. Thus a linear increase in the I-entropy

$$\Delta S_I \simeq h \cdot t$$

can at most hold in an interval of length

$$\tau \lesssim S_{eq}/h$$

Compare this with the relation (12.2) of Chapter 12. We can also
give this restriction the form: The state must remain far from
equilibrium in the time interval where the chaotic property shall
be seen. If the system is near equilibrium, then the I-entropy increase can still be linear but with a coefficient proportional to
ε^2. Then the valid form for the restriction on ε is still (8.2).

Lacking a rigorous theory, we can only postulate that (8.1)
is the relevant form of the condition for effective isolation for
some quantum systems, which we may call chaotic, provided that the
duration of the experiments satisfy some bound of the form (8.3)
and that the system is far from equilibrium. It is interesting to
note that for quantum systems the only information which seems relevant for the validity of this postulate is the spectrum of the
Hamiltonian and the operator character of the perturbations.

8.d Effects on the entropy functions

An instability property of the type described above will have a
considerable significance for the concepts introduced in Chapter 6.
When it exists, there can be no effective way of isolating the system from external perturbations for long periods (except near equilibrium). The effect of the loss of information due to the pertur-

bations is to destroy in part the control of the motion of S given by the fields F and hence the efficiency of the work cycles. This effect of the noise can be expressed in terms of the mobility semigroup $T(F,\varepsilon)$ defined by the perturbed dynamics. The strongly dissipative character of the dynamical maps associated with work cycles of long duration will destroy any group character (indicating reversibility) which the unperturbed $T(F)$ may have as a result of recurrences of long periods. The set of states accessible from a given non-equilibrium state will thus change as a result of the noise. The possibility of reaching the states of lowest energy through work cycles of very long duration will be destroyed.

It is sometimes claimed that the effects of the external perturbations can be replaced by slightly changed initial conditions (a discussion is given in [1] § 1.3). This is not the case here, due to the definition of entropy functions, the values of which for a given state depends on the *potential future evolution* of the system. The noise changes the variational equations for the optimal work cycles in a way which penalizes cycles where S is far from equilibrium for long periods, and as a result the values of the entropy functions change.

Recall the notation $S(\rho;F,D)$ for the entropy defined in Chapter 6 using the Hamiltonian work cycles in $\Gamma(F,D)$, where $D = [0,\tau]$ is their allowed duration. Going a bit ahead of events, we let $S(\rho;F,\varepsilon)$ be the entropy defined in Chapter 10, using the perturbed dynamics and work cycles of arbitrary duration.

First, assume that S is effectively closed during D, i.e. that (ε,τ) satisfy an inequality of the form (8.1) (or (8.2) when relevant) such that ΔS_I is insignificant during D. If we can put $\Delta S_I = 0$, then

$$S(\rho;F,\varepsilon) \leq S(\rho;F,D) \tag{8.6}$$

as the set of available work cycles is larger for the LHS. On the other hand, by (1) of the Proposition in Chapter 10

$$S_I(\rho) \leq S(\rho;F,\varepsilon).$$

If F is so large that, for all $\rho \in E_S$

$$S(\rho;F,D) \simeq S_I(\rho)$$

(compare (3) of the Proposition in Chapter 6), then it holds that

$$S(\rho;F,\varepsilon) \simeq S_I(\rho).$$

For such F the increase in the I-entropy induced by the noise is the source of the increase in the thermodynamic entropy. There is no intrinsic irreversibility: S is seen as a perturbed reversible system and the mobility semigroup $T(F,\varepsilon)$ is "almost" a group.

Next, we consider intervals D such that $\Delta S_I(D) > 0$ must be taken into account. Let the perturbed evolution in D be given by

$$\rho(\tau) = T_\varepsilon(\tau)^*[\rho]$$

where $\rho = \rho(0)$. If this evolution is generated by (a part of) an optimal work cycle (in the sense of Chapter 10), then we find from (1) and (2) of the Proposition of Chapter 10 that

$$S_I(\rho(\tau)) = S_I(\rho) + \Delta S_I$$

$$\leq S(\rho(\tau);F,\varepsilon) = S(\rho;F,\varepsilon).$$

Thus, for all $\rho \in E_S$

$$S(\rho;F,\varepsilon) - S_I(\rho) \geq \Delta S_I \qquad (8.7)$$

where ΔS_I is the increase in the I-entropy in any optimal work cycle for ρ. (8.6) gives a bound on ΔS_I: For every D' such that $\Delta S_I(D') = 0$ to sufficient accuracy

$$S_I(\rho) + \Delta S_I \leq S(\rho;F,D').$$

If $\rho \in G(F)$ then $\Delta S_I = 0$. If F is small, then the durations of the optimal work cycles of Chapter 6 may be of the order of the periods of the spontaneous recurrences. If, in addition, ρ is far from equilibrium, then the noise will ruin them completely through a large value of ΔS_I in (8.7).

The irreversibility of the process where a non-equilibrium state ρ is prepared from an equilibrium state $\rho_0 \in G(F)$ can be measured by the increase in the entropy function $S(\cdot;F,\varepsilon)$

$$S(\rho;F,\varepsilon) - S(\rho_0;F,\varepsilon) = S(\rho;F,\varepsilon) - S_I(\rho_0)$$

$$\geq S(\rho;F,\varepsilon) - S_I(\rho),$$

where the last inequality follows from (3.12). The part

$$S_I(\rho) - S_I(\rho_0) \geq 0$$

is due to the perturbations during the preparation, while the part

$$S(\rho;F,\varepsilon) - S_I(\rho) \geq 0$$

is due to the influence of the noise on the optimal work cycles for the non-equilibrium state ρ.

How does $S(\rho;F,\varepsilon)$ depend on ε? If we admit work cycles of arbitrary duration then, by the argument of Chapter 6.b, the almost periodic property of the unperturbed motion will give the limit (where $D = [0,\tau]$)

$$\lim_{\tau \to \infty} S(\rho;F,D) = S_I(\rho),$$

and by (8.6) and (8.1) (or (8.2)) that

$$\lim_{\varepsilon \to 0} S(\rho;F,\varepsilon) = S_I(\rho).$$

However, if a condition like (8.1) has to be satisfied, with an astronomical value of $t = \tau$, for this limit to be approached, then it has no physical significance.

By (3) of the Proposition in Chapter 6, $S(\rho;F,D)$ is a non-increasing function of τ. Assume that it is a very gentle function of τ of the type discussed in Chapter 6.b, only reaching the minimal value $S_I(\rho)$ for an exceedingly large value of τ. If approximate equality holds in (8.6) for a value of ε satisfying a bound of the form (8.1), then $S(\rho;F,\varepsilon)$ will be a gentle function of $- \ln \varepsilon$ which may be approximately constant over the physically significant decades. This entropy function would then be a measure of the irreversibility of the evolution. It would not be equal to the I-entropy, but its dependence on ε would be extremely difficult to detect as long as ε is not too large.

CHAPTER 9.   THERMODYNAMIC LIMIT

The object of applying the thermodynamic (infinite volume) limit
is at least twofold. First, what is actually measured in those
experiments where thermodynamic concepts are relevant are quantities which do not depend on system size, i.e. either intensive
variables or specific values of extensive variables. Secondly, the
recurrences which are such an inconvenient feature of finite system dynamics are generally expected to disappear in the thermodynamic limit, at least for sufficiently high energies. In this chapter I will sketch a bare outline of what this limit could look like
in the present formalism. The mathematical existence of the limits
introduced here is not established, and it is likely to be a much
more complex problem than for the thermodynamic limit of equilibrium states.

In dealing with equilibrium thermodynamics it is often convenient to deal directly with infinite systems, as is done in the
standard algebraic approach [87]. In the present approach, however,
this method is not appropriate. The work processes of interest
here generally involve global operations on the system S which are
not easily described in a quasilocal picture. As an example take
NMR experiments where the applied magnetic field acts in the same
way on every magnetic moment in the sample. The field will induce
transitions between different "macroscopic" states of S, say states
of different temperature. Such states are disjoint in the quasilocal theory, i.e. there are no local operators which can cause
such a transition. We would also like the "approach to equilbrium"
to consist of a global equipartition of energy rather than in local disturbances flying off to infinity (as happens in quasi-free
models). The case of a local action of the fields actually corre-

sponds to the treatment given in Chapter 7.

There is given a finite quantum system $S^{(1)}$ and a sequence of systems $S \equiv \{S^{(N)}\}$, where $S^{(N)}$ consists of N copies of $S^{(1)}$. There is also given an interaction between the constituents of $S^{(N)}$, the same for all N. The boundary conditions may have to be scaled with N in order to obtain a desired homogeneity property. The fields F, which are the same for all N, are assumed to act in the same way on all constituents of $S^{(N)}$ and in a way which does not depend on N. This means that the number of external parameters is N-independent, and consequently the effective control of the system $S^{(N)}$ will decrease when N increases. There is then a family of sets $F \equiv \{F^{(N)}\}$ describing the control of the system S, where $F^{(N)}$ is defined just as F in Chapter 6. As the parameter space is the same for all N, a cyclic change in the fields defines a family of work cycles $\gamma \equiv \{\gamma^{(N)}\}$, each $\gamma^{(N)}$ acting on $S^{(N)}$ in the manner described in Chapter 6. In addition to this scaling of the work cycles with N, there must also be a scaling of the preparation procedures to give, for each N, a well defined initial state, which is not in general a simple tensor product of states for the constituents.

In order to keep the notation short we write e.g. $\rho \equiv \{\rho^{(N)}\}$ $H \equiv \{H^{(N)} \in F^{(N)}\}$. The standard thermodynamic limits for equilibrium states

$$e(\beta,H) = \lim_{N \to \infty} N^{-1} E(\beta,H^{(N)})$$

$$s(\beta,H) = \lim_{N \to \infty} N^{-1} S(\beta,H^{(N)})$$

exist under fairly general conditions [17,87], and we just take for granted that the specific energy is defined for all reproducible states:

$$\rho(H) = \lim_{N \to \infty} N^{-1} \rho^{(N)}(H^{(N)}).$$

The specific work done by S during a cycle γ is defined as

$$w(\gamma,\rho) = \lim_{N \to \infty} N^{-1} W(\gamma^{(N)},\rho^{(N)}) \qquad (9.1)$$

whenever the limit exists. The mathematical existence of the limit clearly depends on a more detailed specification of the dynamics, a difficult task which will not be attempted here.

In the case of equilibrium properties, the thermodynamic limit is expected to give specific observables without fluctuations, except for some values of the parameters where phase transitions occur. When phase transitions are possible the thermodynamic limit may depend on the limiting procedure, as there can be critical fluctuations of a macroscopic scale [61]. The lack of fluctuations in the general case is due to the physical quantities being of the form

$$X = \lim_{N \to \infty} N^{-1}\sum_1^N X^{(i)} \qquad (9.2)$$

where the $X^{(i)}$ are statistically independent random variables. The central limit theorem then implies that X has no fluctuations [88] In the critical cases the $X^{(i)}$ are correlated.

When the system is driven far from equilibrium with the expenditure of work the situation is more complex. Instability properties like bifurcation phenomena are likely to be important ingredients of the evolution, leading eventually to turbulent behaviour, i.e. with large fluctuations. This transition can not (yet) be treated by a theory which starts from the microscopic dynamics, but the non-linear macroscopic equations of hydrodynamics for driven systems show such a transition to chaotic motion when the strength of the driving is increased [89]. Due to the expected large fluctuations the limiting procedure is certainly very delicate, if at all relevant. The fluctuations are likely to be small only if the irreversible phenomena are spatially homogenous, giving observables of the form (9.2) as a sum of independent contributions. Phenomena which depend on the scale and form of the boundary of the system (like turbulence, convection etc) or involve a strong interaction of the whole system have to be dealt with in terms of finite systems in the present formalism.

Provided that the limit (9.1) exists, the entropy can now be defined in analogy with Chapter 6. To every initial state ρ and work cycle γ there is a final entropy

$$s(\rho;\gamma) = \int_{e_0}^{e(\gamma)} du \; \beta(u,H)$$

where $H = O(\gamma)$, $e_0 = e(\infty,H)$ and $e(\gamma) = \rho(H) - w(\gamma,\rho)$. The *specific thermodynamic entropy* is then defined as

$$s(\rho;F) = \inf\{s(\rho;\gamma); \; \gamma \in \Gamma(F)\} \tag{9.3}$$

The properties of (9.3) follow as in Chapter 6. It should be noted that the standard specific thermodynamic entropy, defined as the limiting specific I-entropy of an equilibrium state, is an affine state function, while the functions (9.3) are concave. This is one aspect of the non-identity of the entropy (9.3) with the I-entropy which will be commented on again at the end of Chapter 11.

The thermodynamic limit is expected to remove most of the problems due to the almost periodic motion of finite quantum systems. The limiting procedure suggested here, where the limit $N \to \infty$ is taken for a fixed work cycle, precludes the exploitation of recurrences which depend on N. Consequently, we can expect $s(\rho(t);F)$ not to be trivially constant in time, even without the introduction of external noise or an upper bound on the duration of the work cycles. It should be stressed, however, that little is known about the recurrence periods as functions of the size of the system. Furthermore, what is known deals exclusively with the *spontaneous* recurrences, i.e. those where the choice of the time-dependence of the fields is *not* exploited to induced a return to the initial state.

In the classical case analytical estimates exist of the recurrence periods of systems of coupled harmonic oscillators [90, 91]. The recurrence properties of a Hamiltonian quantum system with a discrete energy spectrum is not radically different. The return of a density operator

$$\rho(t) = \sum_{k,l=1}^{n} \rho_{kl} \exp(i\omega_{kl}t)$$

near the initial value involves a recurrence condition of the form (where $|\cdot|$ is the trace norm)

$$|\rho(t) - \rho(0)| < \delta. \tag{9.4}$$

The set of values of t which satisfy (9.4) depends on the structure of the spectrum of H in a very complicated way. If the eigenvalues are commensurable, then the motion is strictly periodic. If we assume that there are no rational relations among them at all, then an estimate can be given of the average recurrence period $\tau(\delta)$ defined through

$$\tau(\delta) = \lim_{t \to \infty} t \, r(t)^{-1}$$

where $r(t)$ is the number of recurrences in $(0,t)$ (i.e. the number of intervals in which (9.4) is satisfied). A rough estimate is given in [17] §1.1. A more refined version, with a nice geometric interpretation, was derived by Peres [92]. The asymptotic form, as n (the number of involved energy levels) goes to ∞, is given by

$$\lim_{n \to \infty} \frac{2}{n} \ln[\omega(n)\tau(n,\delta)] + \ln\delta = \text{constant} \tag{9.5}$$

where $\omega(n)$ is an rms average of the $\{\omega_{kl}\}$. Note that the arbitrary phase in the wave functions, which was left out in [92], does not change this form, except that the average ω rather than the average of the energy should be used.

If there are rational relations among the energy eigenvalues, then one can take this into consideration by replacing n by the number of rationally independent eigenvalues in (9.5). It is much more difficult to say anything about the periods of first recurrence, as these will depend on the details of the structure of the spectrum. Regularities, like the existence of approximate rational relations, can lead to a very non-uniform distribution of the recurrences in time. In fact, nothing seems to be known about this problem.

In order to see how $\tau(n,\delta)$ increases with the size (N) of the system $S^{(N)}$, we must estimate how the number of energy levels increases with N. This is easily done by noting that $n(N)$ should be of the order of the number of significantly occupied levels in the equilibrium state of the given mean energy (i.e. of a given value of β). If there is no degeneracy in the spectrum, this number can be found from the equilibrium entropy: With the notations used above

$$S(\beta,H^{(N)}) \simeq \ln[n(N)]$$

This means that for systems without symmetries to cause degeneracies, the number of involved energy levels is expected to increase exponentially with N:

$$n(N) \simeq \exp[N \cdot s(\beta,H)]$$

As the energy range increases at most linearly in N, the density of energy levels also increases exponentially. Finally $\tau(n,\delta)$ increases as a double exponential in N, as $\omega(n)$ is slowly (at most linearly) increasing. This estimate explains why even small systems (like molecules) can have enormous recurrence periods.

CHAPTER 10. THERMODYNAMIC ENTROPY

The threads from the preceding chapters will now be gathered together. Again the quantum system S will be assumed to be finite.
If the thermodynamic limits of Chapter 9 exist, then we can deal
in a similar way with the specific thermodynamic quantities introduced there. External fields F act on S, and this action is described (as in Chapter 2) by a set F of Hamiltonians. There is
also a set of heat baths $R(\beta)$ of the type described in Chapter 4,
one for each value of $\beta \in (0,\infty)$. The other parts (X) of the world
outside S may perturb S in the way described in Chapter 8, but the
energy transfer is assumed to be insignificant. The possibility is
admitted that S can be decomposed into a finite number of subsystems $S = \sum_k S_k$ (which means that the Hilbert space for S is $H =$
$\otimes H_k$), where each S_k can be coupled weakly to any one of the heat
baths in the way described in Chapters 4 and 5.

10.a Thermodynamic processes and entropy

The set F and the coupling of S to a succession of heat baths define a set of thermodynamic processes $P = P(F)$. They include the
work cycles of Chapter 2, cycles composed of the reversible processes of Chapter 5, as well as combinations of these types with
due consideration of the time rescaling of the WCL. Assume that
the elements (cycles) $\gamma \in P$ each have an origin corresponding to
a closed system S. The set P has a composition operation which is
defined as in Chapter 2. Any limitation on the duration of the
processes on the lines described in Chapter 6 is implicitly included in the symbol P. If F defines cycles for which the subsystems S_k are non-interacting, then the reversible processes may be
applied to each S_k separately.

The dynamics of S induced by cycles in P is assumed to be Markovian for the reasons given in Chapter 7. Consequently the correlations between S and the heat baths are neglected. This is most easily justified if we restrict the processes to the composition of a cycle of the type of Chapters 2 and 6 with a subsequent reversible cycle. However, it seems preferrable to formulate the statements as generally as possible. The dynamics can then be of any of the types discussed previously, with the proviso that the processes are such that the thermodynamic inconsistency of the non-Hamiltonian dynamics of open systems, described in section 7.c, is unimportant.

The set P defines a semigroup of dynamical maps on the state space of $S + \sum_\beta R(\beta)$. By the Markov assumption every such map has the property that (compare (7.3))

$$\Delta S_I^S + \sum_\beta \beta \Delta E^{R(\beta)} \geq 0 \tag{10.1}$$

with equality for the reversible processes of Chapter 5. We can now define *work cycles* as those processes in P which are *adiabatic* in the sense that $\Delta S^R = 0$ for each R. Note that this definition depends on the chosen initial state, as ΔS^R is an expectation value: If $\Delta S^R = 0$ for the initial state $\rho = p\rho_1 + (1 - p)\rho_0$, then this need not hold for the initial states ρ_0 and ρ_1. The work cycles refer to the work performed by S, and excludes the consideration of work extracted from the heat baths e.g. by Carnot cycles (on which there is obviously no bound). *Optimal* work cycles are then defined (as in Chapter 6) to be those (of given origin) which lead to the lowest energy state of S.

Using these work cycles, the set $\Omega(\rho;P)$ of states of S accessible from a given $\rho \in E_S$ (without a permanent change in the reservoirs) is well-defined, and so is the set $E(P)$ of reproducible states. As the notion of a work cycle depends on the initial state, there are in general no corresponding dynamical maps which act in an affine way on E_S. A mobility semigroup of dynamical maps can only be defined for those processes where the energy exchange between S and the reservoirs (and X) is zero for all initial states. This mobility semigroup will be essentially $T(F)$ of Chapter 6 or $T(F,\varepsilon)$ of Chapter 8.

The observed quantities in this formalism are the ensemble averages of the work done by S (or each S_k), and of the energy transfer to each $R(\beta)$, during each process in P (i.e. in each time interval). The observation of fluctuations will be dealt with in Chapter 11.

The *equilibrium* states for S are by definition the Gibbs states $G(F)$, which are prepared by coupling all S to one of the heat baths. This means that the states where each S_k is in a Gibbs state with parameter β_k (not all equal) are counted as non-equilibrium states. The coupling of S to the heat baths permits us to consider the set of Gibbs states accessible from a given non-equilibrium state, using processes in P and taking into account the entropy changes in the heat baths.

The *thermodynamic entropy* of S is now defined as a family of entropy functions, one for each set $P = P(F)$ of thermodynamic processes.

<u>Definition</u>: The P-entropy of the state $\rho \in E(P)$ is

$$S(\rho;P) = \inf_\gamma \{S_I(\rho(\gamma)) + \sum_R \Delta S^R(\rho;\gamma)\} \qquad (10.2)$$

where the infimum is over all $\gamma \in P(F)$ leading to final states $\rho(\gamma) \in G(F)$, and where $\Delta S^R(\rho;\gamma) = \beta(R)\Delta E^R$ is the entropy change in R due to the process γ, given the initial state ρ of S.

In order to prove the desired properties of these entropy functions, some restrictions on the set F are necessary. The restriction on F assumed in Chapter 3.b or in (6.12) implies that for every $H \in F$, $\rho \in E(F)$

$$(0,S(\rho;P)) \subseteq \{S(\beta,H); \ \beta \in (0,\infty)\}. \qquad (10.3)$$

This follows from (6.15) and the existence of the Hamiltonian processes in $\Gamma(F,H)$ where $\Delta S^R = 0$. If equality holds, then all $\gamma \in \Gamma(F,H)$ must be optimal. Hence we must have in non-trivial cases

$$\lim_{\beta \to 0} S(\beta,H) > S(\rho;P)$$

for all $H \in F$. Furthermore, F is assumed to be large enough to make

$$\Delta(\beta) \equiv \{S(\beta,H); \ H \in F\} \qquad (10.4)$$

an interval of length > 0 for all $\beta \in (0,\infty)$ (this length will necessarily go to zero as $\beta \to \infty$). From this assumption follows that we can perform reversible Carnot cycles over the whole range $\beta \in (0,\infty)$, using reversible isothermal and adiabatic processes.

10.b Properties of the entropy functions

Proposition: The P-entropy has the following properties.

(1) $S_I(\rho) \leq S(\rho;P(F))$

for all $\rho \in E(F)$, equality holds for $\rho \in G(F)$.

(2) If $\{\rho(t)\}$ is the evolution of the state defined by a $\gamma \in P$, then, for all $s \leq t$,

$$S(\rho(s);P) \leq S(\rho(t);P) + \sum_R \Delta S^R(s,t).$$

Every isentropic process (where equality holds) is part of an optimal work cycle, and every optimal work cycle in P defines an isentropic evolution. The available work $A(\rho;H)$ in cycles of origin H is related to the entropy through

$$S(\rho;P) = \int_0^Q du \ \beta(u,H) \qquad (10.5)$$

where $Q = \rho(H) - A(\rho;H)$ and where we have assumed that $E_0(H) = 0$ (see Chapter 3.b).

(3) $S(\rho;P) \leq S(\rho;P')$ for $P' \subseteq P$.

If $P(F)$ is large enough, such that the final states of (10.2) are accessible through processes which leave the total I-entropy (almost) constant, then (approximately)

$$S(\rho;P) = S_I(\rho).$$

(4) For any set $\{\rho_k \in E(F), \lambda_k \geq 0; \sum_k \lambda_k = 1\}$ it holds that

$$\sum_k \lambda_k S(\rho_k;P) \leq S(\sum_k \lambda_k \rho_k;P).$$

(5) Let $S = S_1 + S_2$ and assume that F does not contain elements with an interaction between S_1 and S_2: $F = F_1 \otimes I_2 + I_1 \otimes F_2$. If, furthermore, F is a sum of two independent parts, $F = F_1 + F_2$, where F_1, F_2 act on S_1 and S_2, respectively, then

$$P \equiv P(F) = P_1 \otimes P_2 \equiv P(F_1) \otimes P(F_2)$$

$$S(\rho;P) = S(\rho_1;P_1) + S(\rho_2;P_2),$$

where ρ_1, ρ_2 are the partial states of S_1 and S_2, respectively. This holds for any state of S , which need not be of the tensor product form $\rho = \rho_1 \otimes \rho_2$.

Proof: (1) From (10.2) follows that for every $\varepsilon > 0$ there is a $\gamma \in P$ such that

$$S_I(\rho) + \Delta S_I^S(\rho;\gamma) + \sum_R \Delta S^R(\rho;\gamma) \leq S(\rho;P) + \varepsilon. \tag{10.6}$$

By (10.1) the LHS $\geq S_I(\rho)$. If $\rho \in G(F)$, then, by (10.2), $S(\rho;P) \leq \leq S_I(\rho)$.

(2) For every $\varepsilon > 0$ there is a $\gamma_1 \in P$ satisfying (10.6) with $\rho \to \rho(t)$, $\gamma \to \gamma_1$. There is no restriction in taking $D(\gamma) = [s,t]$.

Then $\gamma * \gamma_1$ satisfies, by (10.2) and (10.6),

$$S(\rho(s);P) \leq S_I(\rho(s)) + \Delta S_I^S(\rho(s);\gamma * \gamma_1) + \sum_R \Delta S^R(\rho(s);\gamma * \gamma_1)$$

$$= S_I(\rho(t)) + \Delta S_I^S(\rho(t);\gamma_1) +$$

$$+ \sum_R [\Delta S^R(\rho(s);\gamma) + \Delta S^R(\rho(t);\gamma_1)]$$

$$\leq S(\rho(t);P) + \varepsilon + \sum_R \Delta S^R(\rho(s);\gamma).$$

This shows the first statement. In order to prove the second statement, we show that for every γ satisfying (10.6) there is a $\gamma_2 = \gamma' * \gamma_1$, where $\gamma' \simeq \gamma$, such that $\Delta S^R(\rho;\gamma_2) = 0$ for all R. Choose ε, H such that

$$\sup_\beta S(\beta,H) > S(\rho;P) + \varepsilon.$$

Then it follows from (10.4) that a reversible Carnot cycle γ_1 of origin H can be found which transfers the energies

$$\Delta E^R = \beta(R)^{-1} \Delta S^R(\rho;\gamma)$$

from R to S, for every R. Then γ_2 is adiabatic and S ends up with the entropy

$$S_I(\rho(\gamma_2)) = S_I(\rho(\gamma)) + \sum_R \Delta S^R(\rho;\gamma).$$

Consequently, the infimum in (10.2) can be restricted to adiabatic processes where the final state of S is a Gibbs state $\rho(\beta;H)$ such that $S(\rho;P) = S(\beta,H)$. A sequence $\{\gamma_n\}$ of adiabatic processes realizing the infimum corresponds to an optimal work cycle (of origin H) by (6.15). The third statement follows from (3.7) and the equality

$$E(\beta,H) = \rho(H) - A(\rho;H).$$

(3) The first statement is obvious from (10.2). The second is more or less the same as (3) of the Proposition in Chapter 6. From (1)

100

above follows that it is enough, with the present definition, that one state in $G(F)$ is acessible from ρ through a P-process which leaves S_I invariant. If the external perturbations of Chapter 8 can not be neglected, then this state must be accessible by a cycle of such short duration that the I-entropy increase due to the perturbations is negligible.

(4) Let γ satisfy (10.6). If the initial state is ρ_k, then the process γ still gives the final equilibrium state $\rho(\gamma)$ for S (it is defined by a $H \in F$ and a $\beta = \beta(R)$), but the value of $\Delta S^R(\rho_k;\gamma)$ is different. From (10.2) follows that

$$S(\rho_k;P) \leq S_I(\rho(\gamma)) + \sum_R \Delta S^R(\rho_k;\gamma).$$

But the energies $\Delta E^R(\rho;\gamma)$ are affine functions of ρ. Hence

$$\sum_k \lambda_k \Delta S^R(\rho_k;\gamma) = \Delta S^R(\rho;\gamma),$$

and consequently

$$\sum_k \lambda_k S(\rho_k;P) \leq S_I(\rho(\gamma)) + \sum_R \Delta S^R(\rho;\gamma) \leq S(\rho;P) + \varepsilon.$$

(5) The elements of F are, by assumption, of the form

$$H = H_1 \otimes I_2 + I_1 \otimes H_2 .$$

Consequently, the reproducible states do not contain correlations between S_1 and S_2. Note that there can still be an energy exchange between the two systems through the heat baths. Thus, if they are in Gibbs states of different temperature, work can be extracted from S through reversible work cycles. Let β_1, β_2 be such that

$$S(\rho_1;P) = S(\beta_1,H_1), \qquad S(\rho_2;P) = S(\beta_2,H_2).$$

There is then an equilibrium state $\rho(\beta,H)$ which is accessible from $\rho(\beta_1,H_1) \otimes \rho(\beta_2,H_2)$ through S_I-preserving processes, i.e.

$$S(\rho;P) \leq S(\beta,H) = S(\beta_1,H_1) + S(\beta_2,H_2).$$

On the other hand, for any $\gamma \in F$ acting on S, we can decompose the final state entropy and the energy changes in the reservoirs into additive contributions from S_1 and S_2:

$$S_I(\rho(\gamma)) + \sum_R \Delta S^R(\rho;\gamma) =$$

$$= S_I(\rho_1(\gamma)) + S_I(\rho_2(\gamma)) + \sum_R [\Delta S^R(\rho_1;\gamma) + \Delta S^R(\rho_2;\gamma)]$$

$$\geq S(\rho_1;P) + S(\rho_2;P)$$

If γ satisfies (10.6), then the last expression is smaller than $S(\rho;P) + \varepsilon$. This concludes the proof of the Proposition.

It is important to note that the additivity property (5) will not hold if we allow an interaction between S_1 and S_2, even if the state is of the tensor product form $\rho = \rho_1 \otimes \rho_2$. In fact, for interacting systems there can in general be no separate entropy for S_1 and S_2, as the set of accessible final states for S_1 will depend on the initial state of S_2, and vice versa. In other words, the partial states do not not give Markov descriptions for S_1 and S_2 separately in the interacting case.

If there are subsets of processes P_1, $P_2 \subset P$ acting on S_1 and S_2, respectively, and involving no interaction between them, then the entropy functions $S(\rho_1;P_1)$ and $S(\rho_2;P_2)$ can be defined. One may then ask if there can exist a subadditivity relation analogous to (3.3). The answer is negative in general. It is true that if the fields F can be used to control the interaction of S_1 and S_2, while still acting independently on the two systems, in such a way that $P \supset P_1 \otimes P_2$, then it follows from property (3) that

$$S(\rho;P) \leq S(\rho_1;P_1) + S(\rho_2;P_2).$$

On the other hand, if there is no interaction, while F acts on S_1 and S_2 simultaneously, in such a way that the elements in P_1 and P_2 can not be chosen independently, then $P \subset P_1 \otimes P_2$ and it may happen that

$$S(\rho;P) > S(\rho_1;P_1) + S(\rho_2;P_2).$$

This is a likely case if the two subsystems are not separated into macroscopically disjoint bodies.

The formalism used in this chapter has a slight blemish, namely that the definition of an adiabatic process depends on the initial state, and so does the set of allowed work cycles. A consequence of this fact is that it is <u>not</u> generally true that

$$A(\textstyle\sum_k \lambda_k \rho_k;H) \leq \sum_k \lambda_k A(\rho_k;H)$$

as one would expect by analogy with (6.5).

The most important property of the thermodynamic entropy defined above is the fact that there is not one but a family of such functions, one for each set of processes. At first sight, this seems disturbing in view of the unicity of the entropy of classical thermodynamics (if the third law is assumed). However, it was noted already in (1) above that the P(F)-entropy of a state in G(F) is uniquely the I-entropy, which can be identified with the classical entropy. If we increase our control of the system by choosing an F' $\supset$ F, then by (3) the P-entropy of the states in G(F) does not change. Thus, the dependence of the entropy of an equilibrium state on P(F) is implicit through the relation $\rho \in$ G(F). The optimal work cycles associated with these states are of a universal type, namely the Carnot cycles. Note that in order to find these, we do not have to solve the equations of motion. Increasing P by choosing a larger F' $\supset$ F does not change the optimal work cycles for the states in G(F).

Now, let $S(\rho;P(F)) = S_I(\rho)$ for a reproducible state ρ not in G(F). We assume for the moment that the dynamics of S is Hamiltonian (when it is not coupled to the heat baths) i.e. the perturbations of Chapter 8 are neglected. Let ρ be prepared from an initial equilibrium state ρ_0 by action of the fields. Then $S_I(\rho) = S_I(\rho_0)$. But, by assumption

$$S_I(\rho) = S(\rho;P) = S_I(\rho_1),$$

where ρ_1 is the final state defined by the optimal adiabatic process in (2) of the Proposition. Consequently,

$$S_I(\rho_0) = S_I(\rho_1)$$

for the initial and final equilibrium states, and the whole process of preparing ρ and bringing the system back to equilibrium in the optimal way leaves the total system $S + \sum_\beta R(\beta)$ with unchanged entropy. There is a permanent change in this system, and consequently a genuine irreversibility, only when

$$S(\rho;P) - S_I(\rho) > 0 \tag{10.7}$$

This difference is the smallest total increase in the entropy of the world in a process which leads from one equilibrium state to another and involves the preparation of the non-equilibrium state ρ.

Of all the P-entropy functions, the I-entropy has the distinction of being computable without solving the dynamics of the system (for all choices of time-dependent fields!). This is the origin of its predictive value for equilibrium thermodynamics. It allows us to say which equilibrium states are accessible from a given equilibrium state as soon as the entropy of the Gibbs states can be calculated. The price to pay is that the I-entropy is constant in time for a closed system. The non-trivial P-entropies describe the irreversibility due to a limited set of available processes, but they have little predictive value. Note the importance of the fact that the P-entropies are not generally unitarily invariant for the existence of an intrinsic irreversibility.

10.c Irreversibility and approach to equilibrium

The *causes of irreversibility* in the present formalism, described
in detail in Chapters 6-10, can be summed up in the following two
points.

(1) When the mobility semigroup $T(F)$ of S is large enough, such
that $S(\rho;P(F)) \simeq S_I(\rho)$ for all ρ, then the irreversibility is essentially due to the increase in the I-entropy induced by the external perturbations. It can be seen as a decrease in the set of
accessible states resulting from the loss of information content
of the state. Note that we can have an approach to equilibrium,
when equilibrium is defined in terms of the I-entropy of the state
(as it is for $G(F)$), only as a result of the interaction of S with
the rest of the world.

(2) When $T(F)$ is limited, it is not possible, in general, to reach
an equilibrium state of the same I-entropy from a given non-equilibrium state, i.e. (10.7) is significant. The mobility is limited
by a small number of controlled parameters compared to the number
of degrees of freedom of S. It is also reduced by the uncontrolled
perturbations by the external world. When the system is far from
equilibrium, these can be amplified by an instability of the dynamics and restrict the useful work cycles to those of short duration and the reversible ones. The quantity (10.7) increases as a
result of the intrinsic Hamiltonian evolution of S after (or simultaneously with) the driving of S away from equilibrium by the
fields. The increase of the P-entropy is just an aspect of the decrease in the set of accessible states, due to the smallness of
$T(F)$, in every process which takes place far from equilibrium.

One objective of the present formalism is to define a concept
of *approach to equilibrium*. Here "equilibrium" has to be re-defined
in such a way that this relaxation corresponds to an increase in
the P-entropy coming from the intrinsic dynamics of S (defined by

105

a constant $H \in F$, say). A first idea could be to define a state as a P-equilibrium state if it can not be told apart from a Gibbs state of the same energy and P-entropy by applying P-processes to it. However, it is natural to use the property of passivity in P-processes for such a definition. The necessary measure of distance to P-equilibrium is then derived from the P-entropy function itself.

First we observe that a state ρ satisfying

$$\rho(H) = E(\beta,H), \text{ some } H \in F,$$

$$S_I(\rho) < S(\rho;P(F)) = S(\beta,H),$$

is certainly not a Gibbs state due to (1) of the Proposition. On the other hand it shares a significant property with the Gibbs state $\rho(\beta,H)$, namely that of being passive with respect to $P(F)$-cycles of origin H. The passivity follows directly from the relations (3.7) and (10.5) (see property (1) of the relative P-entropy (10.8), below).

The passivity of ρ implies the following property. Let $H_1 \in F$ and put $H_1 - H = \Delta H$. If H_1 is such that $H \pm \delta \Delta H \in F$ for a sufficiently small $\delta > 0$, then

$$\rho(H_1) = \rho(\beta,H)[H_1].$$

In order to show this, define a work cycle $\gamma_\delta = \lim_{\tau \to \infty} \{\gamma_{\delta,\tau}\}$

$$\gamma_{\delta,\tau} = H \text{ for } t = 0,\tau; \quad = H + \delta\Delta H \text{ for } t \in (0,\tau).$$

Let S be coupled weakly to $R(\beta)$ during $(0,\tau)$ and write $\rho_\delta = \rho(\beta,H + \delta\Delta H)$. As described in Chapter 5

$$W(\gamma_\delta,\rho) = \delta(\rho_\delta - \rho)[\Delta H] .$$

The passivity of ρ says that $W(\gamma_\delta,\rho) \leq 0$. The substitution $\delta \to -\delta$ implies that $W(\gamma_{-\delta},\rho) \leq 0$. The linear term in the expansion of

106

$W(\gamma_\delta,\rho)$ in powers of δ must then be zero, i.e.

$$\lim_{\delta \to 0} (\rho_\delta - \rho)[\Delta H] = 0,$$

and this, combined with the same relation where $\rho \to \rho(\beta,H)$, gives

$$(\rho - \rho(\beta;H))[\Delta H] = 0,$$

and the statement follows.

The condition of the statement is true for all $H_1 \in F$ if H is a point in the relative interior of the convex set F. Then the states ρ and $\rho(\beta,H)$ give the same expectations for all elements in F. It is not necessarily true that they give the same values for the work performed in an arbitrary work cycles (recall that the work operators $\hat{W}(\gamma)$ of Chapter 2 are generally not in F). We only know that there is at least one common optimal work cycle for the two states (the reversible one), and that the corresponding values of the work are the same (zero if $O(\gamma) = H$). There is in general a $\gamma \in P(F)$ of origin H such that $W(\gamma,\rho) \neq W(\gamma,\rho(\beta,H))$, and then at least one of these quantities is negative. This means that by performing work on S, it will be possible to tell the states apart. In the case where the optimal work cycle is uniquely the reversible one, then part of this work will be lost, irreversibly.

The question is then if the two states must be considered to be macroscopically different (the words micro/macroscopic are used here in an imprecise, intuitive sense). It was already claimed in Chapter 9 that a system driven far from equilibrium can develop fluctuations on a macroscopic scale. In Chapter 11 it will be argued that microscopic fluctuations can be amplified to a macroscopic level by an instability of the motion. The conclusion which I would like to draw at this point is that it is unreasonable to define a thermodynamic equivalence of states in such a way that they must give the same prediction for cycles performing work on

S in an irreversible fashion. Instead a P-equivalence relation is introduced in the following way. Define a relative P-entropy

$$S(\rho|\rho(\beta,H);P) = S(\beta,H) - S(\rho;P) + \beta[\rho(H) - E(\beta,H)] \quad (10.8)$$

as a distance between ρ and the Gibbs state $\rho(\beta,H)$. Its properties follow from the Proposition above. Only two of them will be given:

(1) $S(\rho|\rho(\beta,H);P) \geq 0$

For finite systems (without phase transitions) equality hold if and only if

$$\rho(H) = E(\beta,H), \qquad\qquad S(\rho;P) = S(\beta,H).$$

These are precisely the states of the given energy which are passive with respect to P-processes of origin H, i.e. which satisfy

$$\rho(H) = E(\beta,H), \qquad\qquad A(\rho;H) = 0$$

(2) $S(\rho(t)|\rho(\beta,H);P)$ is a non-increasing function of t under the evolution defined by H.

The inequality (1) follows from (3.7) and (10.5) which enable us to write (10.8) as (P is fixed)

$$S(\rho|\rho(\beta,H)) = \int_Q^E du[\beta(u,H) - \beta] + \beta A(\rho;H)$$

$$E \equiv E(\beta,H), \qquad\qquad Q \equiv \rho(H) - A(\rho;H).$$

Both terms on the RHS are non-negative. If the LHS = 0, then $A(\rho;H) = 0$ and $E(\beta,H) = \rho(H)$, unless $\frac{d\beta}{du} = 0$. Consequently $S(\rho;P) = S(\beta,H)$ from (10.8).

(2) follows directly from the non-decrease of the P-entropy and the constancy of $\rho(t)[H]$.

The properties of the relative P-entropy suggest that it can be used to define the notion of P-equilibrium states.

<u>Definition</u>: The displacement of ρ from P(F)-equilibrium is

$$d(\rho|G(F)) = \inf\{S(\rho|\mu;P(F)); \mu \in G(F)\}$$

and ρ is called a P(F)-equilibrium state if $d(\rho|G(F)) = 0$. If the infimum is achieved for $\mu = \rho(\beta;H)$, then the function $d(\rho(t)|G(F))$ is non-increasing in t under the evolution defined by H.

In this way a concept of relaxation to equilibrium has been defined, but no general property of relaxation to P-equilibrium can be proved. This may seem to make the formalism sterile, but it is doubtful if it makes much sense to look for such a property. The only problem which seems significant is to find a finite rate of relaxation to P-equilibrium. If there is no such non-zero rate, then the equilibrium state of S will be defined by the properties of the environment of S. In order to have the desirable property that the equilibrium state is determined by the given value of the total energy of S, the relaxation rate must be large enough to bring the relative entropy (10.8) to a value near zero in a time much shorter than the allowed duration of the P-processes (during which S is energetically closed). Clearly the calculation of such a rate is a problem which requires a detailed and realistic model. An abstract scheme is not enough for this purpose.

CHAPTER 11. MEASUREMENTS, ENTROPY AND WORK

We have seen in the preceding chapters that there is in general no
identity between the thermodynamic entropy, as defined there, and
the information-theoretic entropy. The only case where identity
holds is in the limit where there is no intrinsic irreversibility.
This includes the WCL limit for an open system which is always
near equilibrium, and where the irreversibility is due exclusively
to the coupling with the heat bath.

In view of the often repeated identification of entropy with
lack of information, it is interesting to see what happens with
the thermodynamic entropy when we increase our information about
the state of the system through observations. This problem leads
to some comments on Szilard's well-informed heat engine and on the
definition of work for microscopic systems.

11.a Observations on the system

Information-theoretic entropy concepts are related to observations
or sequences of observations made on the system S. We have already
mentioned the Kolmogorov-Sinai entropy (for classical systems),
which refers to the dynamics of S for all time. In the present context the interest lies in the entropy of a state of S at a given
instant and its time dependence. The standard coarse-grained I-
entropy defined by an incomplete measurement on the system at one
instant has the drawback of lacking a non-decrease property (for
finite quantum systems it is almost periodic). Goldstein and
Penrose [29] introduced an entropy function related to all possible
future observations on the system, and having a non-decrease property for basically the same reason as in the case of the P-entropy
defined above. The two entropy concepts are not directly related,

however. The P-entropy deals only with mean values of the observables. Furthermore, in [29] only one fixed dynamics is considered (this leads to problems of physical interpretation which will be discussed in Chapter 12).

At first sight the information-theoretic entropy functions refer to a type of experiment which is completely different from that considered in this work. Typical examples would be photon counting experiments and microscopic observations of Brownian motion. In these instances the outcome of a sequence of observations is well described by a stochastic process. This contrasts with the deterministic behaviour of macroscopic observables (described by ensemble averages), where observations can follow the motion but give no new information not contained in the initial state. There is, however, no clear distinction between the two types of behaviour. In fact, macroscopic indeterministic motion like turbulence is possible, and it is indeed likely, when the system is forced far from equilibrium by the fields. In this type of motion microscopic information is amplified to macroscopic level by the instability of the dynamics. A vivid description of this process has been given by Shaw [72], but mainly referring to non-conservative classical dynamics. A measuring instrument detecting microscopic events is an obvious example. There a sensitive initial state of the instrument can develop into different macroscopic final states depending on the initial microscopic information. The possible appearance of superpositions of different macroscopic states during the evolution poses the question: How will an observation on the system, capable of resolving this mixture, change the entropy concept? Note that in this chapter there will be no attempt to define what is meant by a macroscopic observation or state, instead quite general measurements will be allowed.

The description of measurements in terms of thermodynamic concepts has been treated extensively in the literature ([19,59

93-97] is just a very small selection). Only a bare outline can be given here, explaining the modifications due to the new entropy concept. The property of quantum measurements of disturbing the state of the system will not be an essential feature here. The result of a measurement can then be represented by the replacement $\rho \to \rho_k$ with probability p_k , where

$$\rho = \sum_k p_k \rho_k , \qquad\qquad \sum_k p_k = 1.$$

The *information* associated with the measurement is

$$I\{p_k\} = \sum_k h(p_k), \qquad h(p) = - p \ln p .$$

The concavity of the P-entropy implies that the measurement leads to a reduction of the average entropy by an amount

$$S(\rho;P) - \sum_k p_k S(\rho_k;P) \geq 0.$$

For the I-entropy there is also an upper bound on the entropy re-duction. This is given by equation (2.2) of [19] :

$$S_I(\rho) - \sum_k p_k S_I(\rho_k) \leq I\{p_k\} .$$

This inequality can be applied to the equilibrium states, as the P-entropy coincides with the I-entropy for them. If $\rho \in G(F)$ and $P = P(F)$, then

$$S(\rho;P) - \sum_k p_k S(\rho_k;P)$$

$$\leq S_I(\rho) - \sum_k p_k S_I(\rho_k) \leq I\{p_k\}. \qquad\qquad (11.1)$$

It is important to realize that there need not be such a bound if ρ is a non-equilibrium state. The decrease in the average P-entropy is due to the possibility of optimizing the P-processes in the definition (10.2) for each ρ_k separately. If different equilbrium states are chosen for the ρ_k and P is a small set, then it is clear that the difference may be much larger than $I\{p_k\}$.

11.b Information and entropy

The average entropy decrease obtained through the observation also implies the possibility of obtaining work out of a system in equilibrium. If $\rho = \rho(\beta,H)$, then the state is passive with respect to P-cycles of origin H, but the states ρ_k are not, in general. If we consider S as an open system in contact with a heat bath $R(\beta)$, then the average available work is given by (7.7):

$$\textstyle\sum_k p_k A(\rho_k;H) = \beta^{-1}\sum_k p_k S(\rho_k|\rho(\beta,H)) \leq \beta^{-1} I\{p_k\}$$

where, with the notation of Chapter 7 (compare (10.8))

$$0 \leq S(\rho|\rho(\beta,H)) \equiv S(\beta,H) - S(\rho) + \beta[\rho(H) - E(\beta,H)]$$

$$\leq S_I(\rho|\rho(\beta,H)).$$

The idea that such an observation could beat the second law is of course originally due to Maxwell when he hypothesized what was later called *Maxwell's demon*. A large number of authors have attempted to refute this conjecture by claiming that the process of obtaining the information I must be accompanied by an entropy increase in the measuring instrument which is not less than I [19, 59, 93-97]. The earliest treatment of this problem, introducing the relation between entropy and information for an idealized heat engine, was given by Szilard [95].

The validity of the claim stated above must obviously depend on the choice of entropy function for the measuring instrument. Usually only the I-entropy is considered, and this standard argument will be summed up here for completeness. The notation follows that of [19]. The system $S = (1)$ and the measuring apparatus $M = (2)$ are described as quantum systems. They are assumed to be initially in uncorrelated states ρ_1 and ρ_2, i.e. the total system is in the product state $\rho_1 \otimes \rho_2$. The conservation of the I-entropy for $S + M$ during the interaction gives (prime denoting final states, 1,2 the partial states of S and M, respectively)

$$S_I(\rho_1) + S_I(\rho_2) = S_I(\rho'_{12}) \equiv S_I(\rho'_1) + S_I(\rho'_2) - C_{12}(\rho'_{12}).$$

C_{12} is the logarithmic correlation between S and M in the final state. Theorem 2 of [19] says that the information given by the measurement is bounded by C_{12}:

$$I\{p_k\} \leq C_{12}(\rho'_{12}).$$

Let ρ'_{1k} be the final partial state of S corresponding to the instrument reading (k) of probability p_k. Then

$$\rho'_1 = \sum_k p_k \rho'_{1k} \, ,$$

and the inequality (11.1) reads

$$0 \leq S_I(\rho'_1) - \sum_k p_k S_I(\rho'_{1k}) \leq I\{p_k\}.$$

For an ideal classical measurement $\rho_1 = \rho'_1$. For an ideal quantum measurement this does not hold in general, and

$$S_I(\rho'_1) - S_I(\rho_1) \geq 0,$$

but there is still an average entropy reduction relative to the initial state (equation (7.2) of [19])

$$0 \leq S_I(\rho_1) - \sum_k p_k S_I(\rho'_{1k}) =$$

$$= S_I(\rho'_1) - \sum_k p_k S_I(\rho'_{1k}) + S_I(\rho'_2) - S_I(\rho_2) - C_{12}(\rho'_{12})$$

$$\leq I\{p_k\} \leq S_I(\rho'_2) - S_I(\rho_2).$$

We can also write the last inequality as

$$S_I(\rho_1) + S_I(\rho_2) \leq \sum_k p_k S_I(\rho'_{1k}) + S_I(\rho'_2). \tag{11.2}$$

Thus the non-decrease of the total entropy remains valid if we forget the correlations between S and M and instead replace the final state I-entropy of S by a weighted average over the observed sub-ensembles.

Consider the special case where ρ_1 is an equilibrium state with respect to the unperturbed Hamiltonian H of S. If the energy exchange with M is insignificant, then, by the argument of Chapter 3.c , especially equations (3.10),(3.15),

$$S_I(\rho_1') \leq S_I(\rho_1)$$

$$C_{12}(\rho_{12}') \leq S_I(\rho_2') - S_I(\rho_2).$$

If M is also initially in an equilibrium state, then $S_I(\rho_2')$ = = $S_I(\rho_2)$ and there can be no information gain in the interaction of S and M. Consequently, a measurement on an equilibrium state with no significant energy exchange needs a measuring instrument which is in a non-equilibrium *sensitive* state.

The lack of identity between the I-entropy and the P-entropy of Chapter 10, which alone can measure an intrinsic irreversibility, makes the interpretation of the preceding analysis a debatable point. Can the I-entropy increase in M

$$S_I(\rho_2') - S_I(\rho_2) \geq I\{p_k\}, \tag{11.3}$$

compensating the entropy reduction in S, be considered as a thermodynamic quantity, signifying an irreversible evolution in M? This problem can only be resolved by extending the concepts of Chapter 10 to the combined system $S + M$. There will then be a set P_1 of processes acting on S and similar P_2-processes acting on M, especially for preparing the sensitive non-equilibrium states of M. In addition, the interaction $S - M$ must be taken into account in the set of processes P acting on $S + M$, and evidently

$$P \neq P_1 \otimes P_2.$$

The P-entropy for $S + M$ will have the non-decrease property for all dynamical maps generated by P, including the measurement interaction. The P-entropy will in general not be a sum of contributions from S and M (as explained in Chapter 10). Remember that this holds also for uncorrelated states. For this reason the increase in the P-entropy in a specific process can not be divided between S and M, even if it does not involve interactions or correlations between them. The $P_1 \otimes P_2$-entropy is additive under the assumptions of (5) in the Proposition of Chapter 10, but it need not be non-decreasing under an interaction.

The non-additivity of the P-entropy is an important point which can be expressed in a different way: The irreversibility in an interacting system can not be localized to a part of the system, in general. This is easily seen if the interaction creates correlations between different parts, which can not be exploited by the P-processes. The same is true for a non-interacting but correlated system if we can exploit the correlations through observations on the system. As an illustration, try to define the P-entropies of S and M separately after the interaction, in a way which takes into account the correlations in the fashion of (11.2)

$$S: \quad \sum_k p_k S(\rho'_{1k};P_1), \qquad M: \quad S(\rho'_2;P_2).$$

If the subsequent evolution of M obliterates the information contained in the $S - M$ correlations, e.g. by coupling M to one of the heat baths, then the P-entropy of S increases to the value $S(\rho'_1;P_1)$ (apart from the evolution in S itself). Thus there would be an irreversibility in S due completely to the evolution in $M + \sum_\beta R(\beta)$. Consequently, a picture of separate contributions to a P-entropy for $S + M$, which takes into account the interaction and the information contained in the correlations, is simply not possible. Such an additivity is valid only if the partial states of S and M give Markov descriptions for each system separately, or if S and M are in (uncorrelated) equilibrium states.

An additional argument can be made against the interpretation of the difference (11.3) as a thermodynamic quantity signifying an irreversible evolution. The description of the evolution of $S + M$ in terms of the I-entropy is microscopic and essentially reversible. No distinguishing features has been introduced of M as a macroscopic system or of the states of M corresponding to macroscopically different readings of the instrument. In fact, the abstract formalism is essentially symmetric in S and M. The amount of intrinsic irreversibility in M depends on the choice of P_2. However, the description of the action of the measurement on S does not assume an upper bound on the level of description of M, only a lower bound saying that the relevant $S - M$ correlations are observable.

Many authors have described more or less realistic mechanisms for the apparatus and calculated the entropy production in M on a phenomenological level [96,97]. The average entropy increase is always found to be larger than $I\{p_k\}$. It seems that such models must always make an assumption (perhaps implicit) on how to assign the entropy changes to the two systems. This assumption must also prohibit the reversal of the evolution in which S and M interact to create the correlations necessary for a measurement.

The general conclusion to be drawn from the preceding arguments is again the necessity of a Markov description for the applicability of the entropy concept. In the case of the measurement process a Markov description can be obtained if we restrict ourselves to the partial state of S and leave out a description of the state of M. The measurement is then considered as an operation on S at the level of the dynamical maps induced by the P-processes. The non-decrease of the properly defined total entropy of $S + M$ must be taken into account in a formal way through the addition of a term at least equal to $I\{p_k\}$ to the average entropy of S after the measurement. The information can no longer disappear through the evolution in M. Instead the measurement results are permanent data, per definition. The entropy after such a measurement can then be defined as ($P = P_1$)

$$\sum_k [p_k S(\rho'_k;P) + h(p_k)].$$

This quantity can be considerably smaller than $S(\rho;P)$. An enlarged set of processes PM can now be introduced by adding to the P-processes the measurements M performed by the apparatus M. A new PM entropy is then defined as in (10.2) through an optimal combination of P-processes and M-measurements. For a single available measurement (at $t = 0$, say), the PM entropy of the state at $t = 0-$ can be defined as

$$S(\rho;PM) = \inf\{S(\rho;P), \sum_k[p_k S(\rho'_k;P) + h(p_k)]\} \qquad (11.4)$$

This expression can be generalized in an obvious way to the general case of repeated measurements at arbitrary instants. This entropy will clearly satisfy the non-increase property and $S_I(\rho) \leq S(\rho;PM)$ with equality for equilibrium states.

11.c Exchange of work and heat

A few authors have claimed that the connection between entropy and information is spurious, and that the "well-informed heat engine" of Szilard works just as well without any information on the state of S [98,99]. This conclusion is obviously at variance with the point of view taken here, in spite of the difficulties described above. It seems that the crucial point in the gedanken experiments of Szilard and others is to define the notion of *work* properly in a completely microscopic context. This difficulty was also noted by McClare in connection with the problem of treating the work processes of muscle on a molecular level [100]. It will be shown how a definition of the transfer of work between two microscopic systems may be formulated.

Consider again two systems S, M, where M now includes parts capable of exchanging energy with S, as well as possible measuring devices. M can be considered as a kind of automated version of Maxwell's demon. A less esoteric interpretation can perhaps be found in biochemical energy transfer between molecules, where S may represent an energy-rich molecule (like ATP) and M an enzyme (ATPase) capable of recognizing S and exploiting its energy e.g. to perform muscle work.

The coupling between S and M may be time-dependent, but I assume that this does not involve any appreciable energy exchange with the rest of the world. As before there are defined sets of processes P_1, P_2 on S and M, respectively. Together with the S - M interaction they form a set $P \supset P_1 \otimes P_2$ of processes on $S + M$. The interest now centers on the amount of work which can be extracted from S via M through P_2-processes. The basic idea is that a transfer of energy ΔE from S to M is called work if the work available from M through P_2-processes increases by ΔE. As the engine , which uses the possible information about the state of S to perform work, has been included in M, the final state correlations between S and M are no longer essential and can be discarded. Thus the dynamical maps can be restriced to those where both initial and final states are uncorrelated. In this way the partial states of S and M provide a Markov description of $S + R$, by assumption.

Consider an interaction process in $S + M$ giving the dynamical map

$$T^*[\rho_1 \otimes \rho_2] = \rho_1' \otimes \rho_2'.$$

The energy exchange with respect to the unperturbed Hamiltonian $H = H_1 \otimes I_2 + I_1 \otimes H_2$ is given by

$$\Delta E^S = (\rho_1' - \rho_1)[H_1] = - \Delta E^M = (\rho_2 - \rho_2')[H_2].$$

The work $A(\rho_2;H_2)$ available from M through P_2-processes of origin H_2 is given implicitly by (10.5). One possible definition of the work transfer from S to M is

$$W = A(\rho_2';H_2) - A(\rho_2;H_2).$$

However, this quantity does not represent a net gain, as the system M has to be prepared in the state ρ_2 starting from an equilibrium state. Call this state

$$\rho_2^0 = \rho(\beta_0,H_2).$$

Consider the preparation of the state ρ_2 of M, the interaction of S and M, the extraction of work from M through P_2-processes and the final return of M to the equilibrium state ρ_2^0 through coupling it to $R(\beta_0)$. The net available work in the total cyclic process is

$$W = A(\rho_2';H_2) + \rho_2^0(H_2) - \rho_2(H_2)$$

$$= \Delta E - Q \leq A(\rho_2';H_2) - A(\rho_2;H_2) .$$

$$Q \equiv \beta_0^{-1}[S(\rho_2';P_2) - S(\beta_0,H_2)] \geq 0$$

If there is no interaction, i.e. $\rho_2' = \rho_2$, then $W \leq 0$. We call W the transfer of *work* from S to M, and Q the *heat* transfer. If $Q = 0$, i.e. there is no total entropy increase in $M + R(\beta_0)$, then $W = \Delta E^M$: Work is energy transfer without an entropy increase in the recieving system. Note the asymmetry between S and M in the definition.

At first sight there is a possible objection to this definition of work. If S is in an equilibrium state $\rho_1 = \rho(\beta,H_1)$, then one would expect, from the passivity of ρ_1 with respect to P_1-cycles of origin H_1, that there could be no work transfer from S to M. However, with the suggested definition this does not hold.

In the limit where $S(\rho'_2;P_2) = S_I(\rho'_2)$ it follows from (3.10) and (3.15) that

$$\beta\Delta E^M \leq \Delta S_I^M = \beta_0 Q,$$

$$W \leq \Delta E^M(1 - \beta_0^{-1}\beta),$$

with equality if the total I-entropy is constant. This is the familiar form of the second law associated with a heat engine operating between two reservoirs $R(\beta)$ and $R(\beta_0)$. W may be positive when $\beta < \beta_0$. The availability of work in S is due to the fact that the interaction of S and M is a more general process than that occurring when the fields act on S. It involves, at some stage of the process, the creation of correlations between S and M. This happens when S can react back on M, a thing which is excluded for the strictly classical system F. The action of F is a special limiting case of the interaction where there are no changes in the I-entropies of S and the rest of the world. Another special case is provided by the ideal measurement process of the preceding section. A third case consists of the Carnot processes, where M acts as an ideal heat engine. Then the correlations remain insignificant in the limit of an infinitely slow, reversible process, while the effect of the interaction is seen in the energy transfer. For a general interaction we must expect to find aspects of all these cases.

Now consider the situation where we can extract work from S using P_1-processes as well as by using the machine M in the way described above. There is then a thermodynamic entropy for S based on these processes. It is given by an expression similar to (10.2)

$$S(\rho_1;PM) = \inf\{S_I(\rho'_1) + \textstyle\sum_R \Delta S^R\} \tag{11.5}$$

where the infimum is now over all processes, with the restrictions that M starts and ends in the same equilibrium state ρ_2^0 and that

S ends up in an equilibrium state ρ_1'. The addition of the machine M permits a larger set of processes which includes the possibility of creating S - M correlations. Hence $S(\rho_1;PM) \leq S(\rho_1;P_1)$, and the difference may be considerable, as the P_1-processes do not allow correlations between S and F. This difference includes the effects of using measuring instruments to increase the available work, and thus (11.5) extends the definition (11.4). The entropy cost of the measurement is now taken into account in a less formal way. Still, there is no consistent way of having the P-entropy of $S + M$ decomposed into the PM-entropy of S and a contribution depending only on the state of M. In fact, as noted before, the set of accessible final states of S depends on the state of M. In (11.5) this dependence has been removed by the introduction of the infimum over the allowed evolutions of M. This device precludes the simultaneous consideration of the state of M, of course.

A final remark. Let the measurements defining (11.4) refer to a set of macroscopic properties, and consider macroscopic states with well-defined values for these. Then the specific entropy of an infinite system will be an affine function of the macroscopic states, as the term $\sum_k h(p_k)$ vanishes when divided by the infinite volume. Recall that the specific I-entropy for an infinite system is an affine state function for the same reason [87].

CHAPTER 12. OTHER APPROACHES

This chapter is intended to describe the relations and differences between the entropy concept defined in Chapter 10 and some other notions of entropy existing in the literature. Most treatments depart from a time-homogenous dynamics and make the entropy an information theoretic concept. The concepts of available work and accessible states are not defined. It will be argued below that as a result the thermodynamic interpretation is questionable.

The set of entropy functions defined here can not be claimed to be the only possible ones, as there may exist experiments of a type not covered by the present formalism. There are, however, stringent restrictions on the possible choices. Let $S^*(\rho)$ be an alternative entropy function which satisfies the restriction that $S^*(\rho) = S_I(\rho)$ when ρ is an equilibrium state. Let this entropy be applied to some situation where there is a set $P(F)$ of processes defined by the experimental apparatus. In order that the entropy of the final equilibrium state should never be smaller than that of the initial state, it must hold that

$$S^*(\rho) \leq S(\rho;P(F)). \qquad (12.1)$$

In order that S^* shall satisfy the second law, there must therefore exist an implicit restriction on F. It seems impossible to give such a bound, applicable to all experimental situations, which is more "natural" than others and which could define an intrinsic and non-trivial entropy. Indeed, this would set a priori bounds on the ingenuity of experimental physicists. There can plausibly be just one intrinsic entropy, that is S_I, and this function does not provide a measure of irreversibility for a closed system.

It is interesting to compare the argument above with the definition of irreversibility given by Planck in his well-known text
on thermodynamics [101]. It seems that Planck considered the means
at our disposal to restore the initial state as something given
by nature. The point of view taken here is different, namely that
the concept of irreversibility must depend on the state of the art
in our technology. It should be noted, however, that for large
systems the control of the mobility given by F is likely to be exceedingly small. Increasing this by orders of magnitude will not
necessarily change the value of the entropy very much.

The condition (12.1) underlines the difficulty facing any attempt to construct a function S^* starting only from one specific
time-homogenous dynamics. This is a program which has been pursued
by the Brussels group [102-106]. A certain price has to be paid
in such an approach. In dealing with e.g. spin echo experiments,
(12.1) must be circumvented by introducing a compensating entropy
increase in the apparatus which generates the fields F and performs
the spin flips (see figures 14.1-3 of [102]). This assumption leads
to the conclusion that the decay of the polarization in the spin
system is irreversible, as the restoration of the initial state is
matched by the entropy increase in the apparatus. It will be argued
below that such a picture is not well-founded. The formalism presented here does not ascribe a genuine irreversibility to spin systems showing a perfect echo effect, and there is no need for a
compensating entropy increase in F. This view is consistent with
the treatment of the spin echo phenomenon given in [107].

First, a general remark on the relation between entropy and
work. In classical thermodynamics the distinction between heat and
work is obtained through a partition of the environment of S into
sets of thermalized degrees of freedom (heat baths) and a system
F described by a set of macroscopic parameters. The energy exchanged with the heat baths is called heat, that exchanged with F

is called work. This decomposition into work and heat defines the equilibrium entropy function for S through the condition that for reversible processes the entropy change in S is compensated by the entropy change in the heat baths. Thus, there is a tacit but obvious assumption that there is no entropy change in F associated with the changes in the parameters which define the work cycles.

This assumption was taken over in the present formalism, and it is the basis for the definition of the thermodynamic entropy. Indeed, it would be impossible to get anywhere without it. However, the assumption is abandoned if we associate an entropy increase with the action of the fields in the spin-echo experiment. It seems impossible to accept that such an entropy increase has any fundamental importance without destroying the foundation of classical thermodynamics. One consequence would be that the passivity property of the equilibrium states no longer follows from the second law applied to S + F. In fact, the set of processes allowed by the second law would increase without any obvious limit. When discussing the entropy changes in S, it is of course necessary to distinguish the action of F on S from measurements of the type described in Chapter 11. We saw there that the passivity of the equilibrium states makes no sense when correlations between S and the external world are allowed. No S - F correlations are created and the action of F on S is to perform unitary transformations. In the spin-echo example these can, in principle, be instantaneously inverted.

The crucial point is that the approach chosen by the Brussels group and many others, which has a long tradition going back to Boltzmann, does not include in a natural way the means we may have of reversing the "irreversible" evolution and the possibility of having different accessible final equilibrium states. This may have been a natural thing a century ago, when there was essentially no possibility of influencing the evolution of relaxation processes by external agents, but it is hardly so today. Even the characterization of the equilibrium states becomes a problem without having the very natural condition of passivity, with its obvious thermo-dynamic interpretation, which demands a time-variable Hamiltonian.

Using a fixed time-homogenous dynamics, one can consider experiments involving repeated observations on the system and a corresponding information-theoretic entropy. The most rigorous result in this direction (for classical systems) is that of Penrose and Goldstein already mentioned [29]. Their construction can be seen as providing a Markov description of the observational process. This is done by introducing an algebra generated by the set of all coarsegrained future observations on the system. A crucial assumption in the definition of the entropy is that the measure on the algebra generated by a non-equilibrium initial state is absolutely continuous relative to that defined by the equilibrium state. Together with the assumption of K-mixing, this implies that the two measures are the same asymptotically (in time), i.e. there is relaxation to equilibrium (compare with the local perturbations of equilibrium states of infinite systems, where there is a corresponding spatial approach to equilibrium [108]). The very strong assumption of the K-mixing property, which removes all problems associated with recurrences, would clearly simplify the present formalism, if it could be applied to Hamiltonian systems.

The K-mixing property is basic also for the recent work by Misra, Prigogine and coworkers [103-106]. They claim that the dynamics of K-systems have an intrinsic irreversibility not due to coarse-graining or other approximation procedures. This philosophy seems to disagree completely with that of the present work. However, I will argue that the application of K-mixing models to physical systems involves an approximation of a coarse-graining nature. The problem is then transferred to that of justifying the model by some more detailed understanding of the physics.

For abstract dynamical systems the concept of energy is not defined. Ergodic systems with a unique stationary probability distribution can instead be thought of as representing the dynamics on a surface of constant energy with a microcanonical equilibrium state. The I-entropy defined by the probability distribution can then represent the deviation of the entropy from the equilibrium value: With the notation of [29]

$$S_{eq} - S = \ln\mu(\Omega) - \int_{\Omega} d\mu \ \rho\ln\rho$$

where $\int d\mu = \mu(\Omega)$, $\int d\mu \ \rho = 1$. Without the introduction of phase space cells (volume $\hbar^N$, where N is the number of degrees of freedom) the equilibrium thermodynamic entropy of a classical Hamiltonian system is undefined. This is even more true of abstract dynamical systems, which represent physical systems of undefined size due to the lack of constants of physical dimension. Consequently, the physical value S_{eq} of the equilibrium entropy must be prescribed in a way which relates the model to the physics. If a given K-shift T (for a K-flow T_t , take $T = T_1$) has a positive KS entropy ($h(T) > 0$), then this dynamics is certainly irreversible in the sense of ergodic theory, but the corresponding physical rate of relaxation must depend on the quantity $h(T)S_{eq}^{-1}$, and hence on the assigned value of S_{eq}.

Due to the finite value of S_{eq} of a physical system, the amount of information that can be obtained by observing the system is limited:

$$I \leq S_I(\rho_{eq}) = S_{eq},$$

provided that the system is considered to be closed and that the perturbations due to quantum measurements can be neglected. It follows directly from the definition of the KS entropy that the information contained in n repetitions of an observation corresponding to a generating partition must satisfy: $n \cdot h(T) \leq I_n$.

The inequality $I_n \leq S_{eq}$ then gives a limit on the validity of the K-mixing model: We will not see the lack of genuine K-mixing asymptotic behaviour (measured by a $h(T) > 0$) only if

$$t \ll S_{eq} h(T)^{-1} \qquad (12.2)$$

For values of t of the order of the RHS the effects of quantum measurements can not be neglected.

If $h(T)$ corresponds to a manageable amount of information from real observations of the system, and if S_{eq} is the entropy of a macroscopic system, then the limit (12.2) provides an enormous time span. On the other hand, one may attempt to identify $h(T)$ with the rate of increase of the thermodynamic entropy (Goldstein showed that it is the asymptotic rate of increase of the Goldstein-Penrose entropy when the chosen partition is fine enough [109]). With this interpretation the model should be valid for a time t such that $t \cdot h(T)$ is of the same order of magnitude as S_{eq}, in order to have a significant increase of the entropy. This contradicts the use of the asymptotic K-mixing property according to (12.2).

The amount of information corresponding to macroscopic values of S_{eq} is so enormous that there is, in general, no relation between the information obtained in measurements on a system and the thermodynamic entropy. An exception is provided by the observation of thermal fluctuations, where the average entropy change due to the measurement is related to the information gain by (11.1). This is due to the fact that the a priori distribution for the outcomes of the measurement is given by the equilibrium state. When the system is forced far from equilibrium (and the a priori distribution is different), the deviation from equilibrium can easily be observed. In fact, if a macroscopic quantity is observed to have a value far from the equilibrium one, we will conclude that S_{eq} - S is considerable, even if the information content of the measurement is insignificant on the thermodynamic scale. The increase of the entropy after such an observation is a transient phenomenon, with no obvious relation to the KS entropy of a K-mixing model. It should be noted that the Goldstein-Penrose entropy can describe also a transient entropy production much larger than h(T).

From the preceding arguments it seems clear that the K-mixing models can be justified for all values of the time parameter only if we can allow the limit $S_{eq} \rightarrow \infty$ ($\hbar \rightarrow 0$) to be taken. Obviously this limit corresponds to a kind of coarse-graining. The physical interpretation is that the observations on the system give an amount of information which is infinitesimal on the scale of S_{eq}. This restriction is fulfilled if we consider only departures from equilibrium in a few degrees of freedom of a large system on the lines described in Chapter 7.

Due to the difficulties of defining a non-trivial and intrinsic entropy function for non-equilibrium states, the idea of a *subjective* or *anthropomorphic* entropy was advocated by many authors. The dependence of the entropy on the type of experiments performed on the system was clearly expressed by Bridgman [110]: "The entropy in general therefore must depend on the universe of operations and must change when the universe changes". The precise nature of this dependence was not described, however. The point of view that there are many entropy functions, each corresponding to a certain level of description, was also taken by Grad [111]. He ascribed the observed irreversibility to the sensitive dependence on initial conditions (see Chapter 8).

The subjective entropy idea was most consistently developed by Jaynes [7,112]. An information theory setting is used, with the entropy measuring the experimenters degree of ignorance of the state of the system. The maximum entropy formalism is thus presented as a form of statistical inference. It allows the construction of an ensemble containing the information that the expectations of a number of observables have given values at certain instants and having the largest I-entropy consistent with this fact. The second law then becomes a statement on the increase of the I-entropy in reproducible experiments which is quite similar to (7.3). However, the maximum entropy formalism does not provide any specific mechanism for the increase of the entropy in a closed system, except for the general statement that the information contained in the past history of the system becomes less relevant as time passes.

The formalism presented here has the advantage of having a definite notion of relevance for the information contained in the state of the system. This relevance is that it can be used to produce work through a given set of work processes. This approach also makes a direct contact with thermodynamics. It seems that the P-entropies of Chapter 12 can not be interpreted as maximum I-entropy functions. If we consider the information contained in the preparation procedure of a reproducible state, then this defines an entropy which is uniquely the I-entropy. In fact, for any given preparation procedure we can, in principle, choose $P(F)$ as large as we like, thus obtaining the I-entropy in the limit. On the other hand, we can consider the predictions about the future which we can make from a given state. Let ρ be a state which is P-equivalent to an equilibrium state $\rho(\beta,H)$ in the sense that the relative P-entropy (10.8) vanishes. Assume that there is a state ρ' equivalent to ρ with respect to all P-processes, i.e.

$$W(\gamma,\rho') = W(\gamma,\rho), \qquad \text{all } \gamma \in P(F),$$

and furthermore that

$$\rho'(H) = \rho(H), \qquad S_I(\rho') = S(\rho;P).$$

Then we find that $S_I(\rho') = S(\beta,H)$, and from (3.4) and (3.5) it follows that

$$\rho' = \rho(\beta, H).$$

But it was indicated in Chapter 10.c that ρ and $\rho(\beta, H)$ need not give the same values of the work for all P-processes, and thus we obtain a contradiction. This means that a state ρ' which gives the same predictions as ρ will satisfy

$$S_I(\rho') < S(\rho; P),$$

in general.

Again, the maximum entropy formalism does not consider the methods we may have of reversing the motions. The only treatments of irreversible thermodynamics which take into account the processes generated by the response of the system to external forces are those based on macroscopic models, like those of Meixner [50] mentioned in Chapter 6.

On a more formal level, the concept of thermodynamic processes can be developed in the context of rational thermodynamics. The erudite reader may have noticed a similarity between the formalism set out in Chapter 10 and the thermodynamics of Coleman and Owen [113] and that of Day [114]. The entropy functions defined in Chapter 10 correspond quite closely to the "largest entropy" of [113] (page 40). If we write

$$s(\rho, \gamma) \equiv - \sum_R \Delta s^R(\rho, \gamma)$$

$$m(\rho, \rho_0) \equiv \sup\{s(\rho, \gamma); \gamma \in P\},$$

where the supremum is over all processes mapping the state   into a reference equilibrium state ρ_0, then this notation corresponds to that of equations (5.6) and (8.6) of [113], and the P-entropy is given by

$$S(\rho; P) = S(\rho_0) - m(\rho, \rho_0).$$

In the rational thermodynamics approach there is, of course, no proper dynamics to give a sense to the idea of irreversibility. Instead, the irreversibility of each thermodynamic process in the scheme has to be postulated. For the same reason, the energy and entropy functions have to be prescribed in order to define the model, which is then of a purely phenomenological nature. In contrast to this situation, the statistical mechanics of Hamiltonian systems provides a direct relation between the energy and entropy functions for equilibrium states through (3.7). In the present formalism this relation is extended to all states through the use of the available work, leading to (10.5). The fundamental property of Hamiltonian systems which makes thermodynamics possible is the multiple roles of the Hamiltonian in defining the dynamics, the concepts of energy and work and the entropy of equilibrium states.

APPENDIX A. QUANTUM MARKOV PROCESSES

A short introduction is given to the topic of irreversible Markovian dynamics in open quantum systems. The problems associated with a thermodynamic interpretation of this type of dynamics are described.

A.1 Reduced dynamics

Consider a small system S interacting with a large reservoir R. The dynamics of $S + R$ defined by a Hamiltonian H_{S+R} generates a family of dynamical maps for S, all depending on the assumed initial state of R:

$$T(t)[X] = Tr_R[\rho_R U(t)^+ [X \otimes I]U(t)], \quad t > 0, \tag{A.1}$$

where $X \in B(H_S)$, $U(t) = \exp[-\frac{i}{\hbar} t \cdot H_{S+R}]$. These maps have a strong positivity property called *complete positivity* (CP for short)

$$\sum_{i,j} Y_i^+ T[X_i^+ X_j] Y_j \geq 0, \tag{A.2}$$

for all $\{X_i, Y_i \in B(H_S)\}_1^n$, all n. This property can also be formulated in the following way: In the tensor product space $B(H_S) \otimes M_n$ of $n \times n$ matrices with operator-valued entries $X = (X_{ij} \in B(H_S))$, T defines a map $T_n = T \otimes Id$,

$$(T_n[X])_{ij} = T[X_{ij}],$$

which is positive (and in fact CP) for all n. Every normal CP map can be written in the form

$$T[X] = \sum_i V_i^+ X V_i, \tag{A.3}$$

where $V_i \in B(H_S)$. Without the CP property the dynamical maps would be inconsistent, in the sense that the dynamics lifted in a trivial way to a larger system containing S would not necessarily preserve the positivity of the states. See [13] for a more detailed background on the properties and physical interpretations of CP maps.

In general the reduced dynamical maps for S do not provide an interesting description, as they depend on the chosen initial state of R at a particular instant in the past. Only with some additional assumptions will there be an autonomous time-homogenous evolution in S. One necessary condition is clearly that the state of R does not change significantly as a result of the interaction with S.

The simplest type of reduced dynamics is a semigroup of dynamical maps

$$T(s + t) = T(t) \cdot T(s). \tag{A.4}$$

The conditions (A.2), (A.4) and an assumption of norm continuity

$$\lim_{t \to 0} \| T(t) - Id \| = 0$$

and of conservation of probability

$$T(t)[I] = I,$$

allows a complete description of the *generator* L of the semigroup:

$$T(t) = \exp(tL), \qquad L = \frac{d}{dt} T(t) \Big|_{t=0} .$$

It can be proved that L has the form

$$L[X] = L_d[X] + \frac{i}{\hbar}[H,X], \tag{A.5}$$

$$L_d[X] = \frac{1}{2} \sum_i [V_i^+, [X, V_i]],$$

where $V_i \in B(H_S)$, $H^+ = H$ [13-15]. The decomposition of L into a *dissipative* part and a Hamiltonian part is not unique in general. However, the dissipation, described by the bilinear expression

134

$$D(X,Y) = L[XY] - L[X]Y - X L[Y],$$

is unique. The CP property of T implies that

$$\sum_{i,j} Y_i^+ D(X_i^+, X_j) Y_j \geq 0. \qquad (A.6)$$

L is Hamiltonian if and only if D = 0.

The evolution of the state of S expressed by the dual maps

$$T(t)^*[\rho] \equiv \rho \cdot T(t)$$

is defined by what is often called a *Markovian master equation*

$$\frac{d}{dt} \rho(t) = L^*[\rho(t)], \qquad (A.7)$$

where L^* has the explicit form

$$L^*[\rho] = \frac{1}{2} \sum_i ([V_i \rho, V_i^+] + [V_i, \rho V_i^+]) - \frac{i}{\hbar} [H, \rho].$$

The detailed justification of such a simple type of evolution is extremely difficult and involves limiting procedures, which, while being mathematically well defined, do not have a very clear physical interpretation. As the evolution of the total system S + R is Hamiltonian and does not define a direction in time, the limiting procedures and approximations which lead to a dissipative evolution must do so. The most used limiting procedure involves the weak coupling limit (=. WCL = van Hove limit) where the strength λ of the interaction between S and R and the time parameter t is scaled in such a way that

$$\lambda \to 0, \quad t \to \infty, \quad \lambda^2 t = \text{constant.}$$

The most thorough treatment of the derivation of the master equation in the WCL limit was performed by E.B. Davies [37-39]. An extensive review of the WCL and similar limiting procedures has been given by Spohn [40].

The case where the reference state of R is a KMS state (parameter β) allows more information to be given on the structure of
the generator: It can be decomposed into a Hamiltonian part (which
corresponds to the unperturbed evolution) and a dissipative part,
such that the two parts commute and L_d^* annihilates the Gibbs state
$\rho(\beta,H)$ (and so does L^*, of course) [24]

$$L_d^*[\rho(\beta,H)] = L^*[\rho(\beta,H)] = 0. \qquad (A.8)$$

Hence this Gibbs state is invariant under the semigroup generated
by L^*. Furthermore, L_d satisfies the relation called *detailed
balance* [24,43,44]: For all $X \in B(H_S)$

$$L_d^*[X \rho(\beta,H)] = L_d[X] \rho(\beta,H).$$

When $\beta = 0$ (infinite temperature) the generator is of a special
form: In (A.5) all the V_i can be chosen to be self-adjoint ($V_i^+ =
V_i$) [64,65]. This means that the dissipative part is of the form
(3.11), which is also that obtained from a stochastic Hamiltonian
perturbation.

A.2 Markov processes

The semigroup evolution of the idealized reduced dynamics in A.1
and the corresponding master equation immediately recalls similar
evolution equations in classical physics, e.g. the Fokker - Planck
equation. It is well known, however, that the semigroup property
(Chapman - Kolmogorov equation) does not suffice to define a corresponding Markov process as e.g. the Ornstein - Uhlenbeck process
[115]. In order to qualify as a Markov process, a random evolution
must have well-defined higher order conditional probabilities
(corresponding to repeated observations of the system) all derived
from the basic transition probability $\{p(x|y)\}$ (which corresponds
to the dynamical map T): If $x_1,\ldots,x_n$ are the values of a random
variable X observed at $t_1 < t_2 <\ldots< t_n$, then

$$p(x_2,\ldots,x_n|x_1) = p(x_n|x_{n-1})p(x_{n-1}|x_{n-2})\ldots p(x_2|x_1).$$

In order to define an analogous property for quantum systems, it is necessary to introduce the action of measuring instruments on the system. A general scheme called quantum stochastic processes was defined in [56]. Such a process defines the probabilities of the outcomes of any sequence of observations on S. The observations are represented by a set of dynamical maps which are often called *operations*. In the present context the full formalism is not required. We just recall that the operations are completely positive maps (called E). Examples are the von Neumann expectations

$$E[X] = \sum_i E_i[X], \qquad E_i[X] = P_i \, X \, P_i,$$

where the P_i are the orthogonal projections of the spectral resolution of some observable.

It is important to realize that the set of operations on a quantum system has a structure which is different from that of the commutative case. The expectations above are not closed under composition and convex combination. On the other hand they are limiting cases of convex combinations of unitary operations. Limits of products of projection-type maps can act as unitary operations on some states. Thus, there is in general no subset of all the CP maps which can be separated out as a set of operations. Especially, there is no absolute separation between the operations corresponding to measurements and the unitary operations which can be considered as describing the action of time-dependent external fields.

The natural way of generalizing the classical Markov property is to define the dynamical map resulting from a sequence of observations on the system as a simple time-ordered composition of the dynamical maps of the semigroup evolution and the operations

$$T(E_1,\ldots,E_n) = T(t_1-t_0)\cdot E_1\cdot T(t_2-t_1)\cdot\ldots T(t_n-t_{n-1})\cdot E_n \quad (A.9)$$

The associated probability, given that the initial state of S at $t_0 < t_1$ is ρ, is [56,57,63]

$$p(E_1,\ldots,E_n;\rho) = \rho(T(E_1,\ldots,E_n)[I]).$$

A justification of the expression (A.9) has recently been given by Dümcke for the WCL and a singular coupling limit [116].

If we now consider S in a time-dependent field, then the resulting evolution can be regarded as a limiting case of (A.9), where the operations are now unitary. The dynamical maps $T(s,t)$: $X(s) \to X(t)$ are defined by the solutions of

$$\frac{d}{dt} X = L(t)[X] \qquad\qquad (A.10)$$

$$L(t) = L_d + \frac{i}{\hbar} [H(t),\cdot],$$

where L_d is a dissipative generator of the form given in (A.5). A direct consequence of this equation is that the dynamical maps have the semigroup property

$$T(s,u) = T(s,t)\cdot T(t,u) \qquad\qquad (A.11)$$

for all choices of $H(t)$. Note that the evolution of the state (described by the dual maps) satisfies the Markov property defined in Chapter 7.a: For all $t > s$

$$\rho_S(t) = T(s,t)^*[\rho_S(s)] \ .$$

The equivalence of (A.9) and (A.10) in a well-defined sense is proved in [57]. In order to see this relation in an intuitive way, let the external fields be δ-functions in time

$$H(t) = H_0 + \sum_k \delta(t - t_k) \, h_k$$

Then the solution of (A.10) is of the form (A.9) with the operations given by

$$E_k[X] = U_k^+ X U_k, \qquad U_k = \exp(\tfrac{i}{\hbar} h_k),$$

$$T(t) = \exp(L_d + \tfrac{i}{\hbar}[H_0, \cdot]) t \ .$$

If we let $H(t)$ be a step function $H(t) = H_k$ for $t \in \Delta_k$, where Δ_k are intervals on R, then the evolution defined by (A.10) in a union of such intervals is again a composition of maps

$$T(\cup \Delta_k) = T(\Delta_1) \cdot T(\Delta_2) \cdot \ldots \ T(\Delta_n)$$

$$T(\Delta_k) = \exp(L_d + \tfrac{i}{\hbar}[H_k, \cdot]) |\Delta_k|$$

The general case is recovered by the Lie - Trotter formula [61], which in this case can be written

$$T(s,t) = \lim_{\varepsilon \to 0} T(\Delta_1) \cdot T(\Delta_2) \cdot \ldots \ T(\Delta_n),$$

where $\cup \Delta_k = (s,t)$, $\sup_k |\Delta_k| \le \varepsilon$.

A consequence of the semigroup property (A.11) is that the solutions of the corresponding master equation for the states

$$\frac{d}{dt} \rho(t) = L_d^*[\rho(t)] - \tfrac{i}{\hbar}[H(t), \rho(t)] \tag{A.12}$$

satisfy a type of H-theorem: For any two initial states ρ, μ the relative entropy of the time-evolved states is non-decreasing in time: For all $s < t$

$$S_I(\rho(t) | \mu(t)) \le S_I(\rho(s) | \mu(s)). \tag{A.13}$$

This follows directly from (3.5) and the semigroup property. The interpretation is that it becomes more difficult to tell the two states apart as time goes by, and the loss of information is mon-otonic. It can be proved that this is a characteristic property of the Markovian type of evolution [25]. For non-Markovian systems there are always choices of $H(t)$ and the initial states ρ, μ such that $S_I(\rho(t) | \mu(t))$ increases in some time interval. This type of

139

behaviour is analogous to the spin echo effect. Clearly there can not be a monotonic decrease property if it is possible to recreate the initial state by choosing a suitable time-dependence for the Hamiltonian.

From the information-theoretic point of view the Markov processes and the loss of information they represent are the most natural descriptions of irreversibility. It is unfortunate that this irreversibility can not be given a thermodynamic interpretation. This will be seen in the next section.

A.3 Non-passivity of Markov processes

Let L_0 be the WCL form of the semigroup generator corresponding to the inverse temperature β of the reservoir and the unperturbed Hamiltonian H_0 for S. Then we write, as in A.1

$$L_0 = L_d + \frac{i}{\hbar}[H_0, \cdot],$$

where L_0^* annihilates $\rho(\beta, H_0)$, by (A.8). The generator obtained by changing the Hamiltonian to $H \neq H_0$ but keeping L_d (in line with (A.12)) will not satisfy (A.8) with the new Gibbs state $\rho(\beta, H)$, in general, but it will have some other invariant state, which is most often unique as argued in Chapter 4. A consequence of this fact is that the dynamical system defined for an arbitrary time-dependent Hamiltonian through (A.12) will not have the passivity property. This means that if the work performed by $S + R$ is defined by (2.7), where $\rho_S(t)$ is the solution of (A.12), then $W(\gamma, \rho) > 0$ for suitable choices of H and γ with $O(\gamma) = H$, when ρ is the invariant state for the value H of the Hamiltonian. It seems quite difficult to show this in the general case, as there is no explicit formula for the invariant state for $H \neq H_0$. It is sufficient, however, to prove the statement in a special but typical case through an explicit calculation. The simplest example is provided by the Bloch equations, which are equivalent to a Markovian master equation in a two-dimensional Hilbert space.

The Bloch equations with homogenous relaxation (rotation symmetric dissipative part) are as follows [59]

$$\frac{d\bar{M}}{dt} = -\lambda(\bar{M} - \bar{M}_0) + g\bar{M} \times \bar{H}(t) \qquad (A.14)$$

where $\bar{H}$ is the applied magnetic field, $\bar{M}$ the magnetic polarization of the sample, λ the inverse of the relaxation time, g the gyromagnetic ratio and $\bar{M}_0$ the equilibrium magnetization in a fixed external field $\bar{H}_0$. $\bar{M}_0$ is given by Curie's formula $\bar{M}_0 = \frac{N}{4}g^2\hbar^2\beta\bar{H}_0$, where N is the number of spins, which we put equal to 1. The corresponding master equation can be found from the relations

$$M(t) = Tr(\rho(t)\bar{\mu}), \qquad H(t) = -\bar{\mu} \cdot \bar{H}(t), \qquad \bar{\mu} = \frac{1}{2}g\hbar\bar{\sigma}$$

where $(\sigma_1, \sigma_2, \sigma_3)$ are the Pauli matrices. The work can now be expressed in the following way

$$W(\gamma, \rho) = \int dt \, \bar{M}(t) \cdot \frac{d\bar{H}(t)}{dt}.$$

Partial integration gives, over the interval $(0, \tau)$

$$W = (\bar{M}(\tau) - \bar{M}(0)) \cdot \bar{H}(0) + \lambda\int_0^\tau dt(\bar{M} - \bar{M}_0) \cdot \bar{H}$$

where $\bar{H}(\tau) = \bar{H}(0)$ has been used. Note that $\bar{H}(0) \neq \bar{H}_0$. Consider the special type of cycle where

$$\bar{H}(t) = \bar{H}(0) \text{ for } t \leq 0, \, t \geq \tau,$$

$$= \bar{H} \quad \text{for } 0 < t < \tau.$$

Then we find that

$$\int_0^\tau dt(\bar{M}(t) - \bar{M}_0) \cdot \bar{H}(t) = \bar{H} \cdot \int_0^\tau dt(\bar{M}(t) - \bar{M}_0)$$

$$= \bar{H} \cdot (\bar{M}(0) - \bar{M}_0)\int_0^\tau dt \, \exp(-\lambda t),$$

$$W = \bar{H}(0) \cdot (\bar{M}(\tau) - \bar{M}(0)) + (1 - e^{-\lambda\tau}) \, \bar{H} \cdot (\bar{M}(0) - \bar{M}_0).$$

The relation between $\bar{H}(0)$ and the stationary value $\bar{M}(0)$ predicted from (A.14) is obtained from

$$- \lambda(\bar{M} - \bar{M}_0) + g\bar{M} \times \bar{H}(0) = \bar{0}, \tag{A.15}$$

and the solution is

$$\bar{M} = \bar{M}(0) = (\lambda^2 + g^2|\bar{H}(0)|^2)^{-1} \cdot$$

$$\cdot [\lambda^2\bar{M}_0 + g\lambda\bar{M}_0 \times \bar{H}(0) + g^2(\bar{M}_0 \cdot \bar{H}(0))\bar{H}(0)],$$

where we have used that $\bar{M}(0) \cdot \bar{H}(0) = \bar{M}_0 \cdot \bar{H}(0)$, which follows directly from (A.15). Also note that $\bar{H}(0) = c\bar{M}_0 \Rightarrow \bar{M}(0) = \bar{M}_0$.

Now let $|\bar{H}| \to \infty$, $\tau \to 0$, in such a way that $g\tau|\bar{H}| = \varphi$ and $\hat{e} = |\bar{H}|^{-1}\bar{H}$ are constant. Then we can neglect the dissipation in the calculation of $\bar{M}(\tau)$:

$$\bar{M}(\tau) \simeq \bar{M}_1(0) + \bar{M}_2(0) \cos\varphi + \bar{M}(0) \times \hat{e} \, \sin\varphi$$

where $\bar{M}_1(0) = (\bar{M}(0) \cdot \hat{e})\hat{e}$, $\bar{M}_2(0) = \bar{M}(0) - \bar{M}_1(0)$, and in the limit we find

$$W = \bar{H}(0) \cdot (\bar{M}(\tau) - \bar{M}(0)) + \lambda\tau\bar{H} \cdot (\bar{M}(0) - \bar{M}_0)$$

$$= - \bar{H}(0) \cdot \bar{M}_0(1 - \cos\varphi) + (\bar{M}(0) \cdot \hat{e})(\bar{H}(0) \cdot \hat{e})(1 - \cos\varphi)$$

$$+ \bar{H}(0) \cdot \bar{M}(0) \times \hat{e} \, \sin\varphi + g^{-1}\lambda\varphi\hat{e} \cdot (\bar{M}(0) - \bar{M}_0)$$

This expression is clearly not negative definite. In fact, choose $\bar{H}(0) = c\bar{M}_0$ ($c < 0$), $\hat{e} \cdot \bar{M}_0 = 0$, $\varphi = \pi$, to find that $\bar{M}(0) = \bar{M}_0$ and $W = 2|\bar{H}(0) \cdot \bar{M}_0|$. Consequently, W can be made as large as we please. As the rate of relaxation to the stationary state defined by $\bar{M}(0)$ is determined by λ, it is evident that the average rate of work which can be extracted from the system is of the order of magnitude of λW. This is also the order of the rate of energy dissipa-

142

pation in (A.14). Hence, even if the absolute size of this discrepancy is small in this system of a single spin $-\frac{1}{2}$ particle, it is true that the model fails to be passive in a maximal way. Note that for work cycles of arbitrary origin we can combine a reversible cycle with one of the form described above to show that the non-passivity always holds. Also note that the preceding arguments fail if $\beta = 0$ (or $\hbar = 0$), as it then follows that $\bar{M}_0 = \bar{0}$. For $\beta > 0$ the conclusion is that (A.14) is not valid for arbitrary $\bar{H}(t)$, and clearly (A.15) does not give the correct stationary states for $\bar{H} \neq \bar{H}_0$.

It may be of some interest to recall that Redfield discovered a different type of inconsistency between the Bloch equations and thermodynamics [117]. He actually found an experimental deviation from the predictions based on (A.14).

A.4 Non-KMS property of Markov processes

The property of non-passivity of quantum Markovian dynamics treated above refers to the response of the system to time-dependent forces. It is interesting to observe that even the time-homogenous dynamics is inconsistent with the properties of thermal equilibrium. This is seen from the fact that the KMS property, which must hold for the exact reduced dynamics, can not be satisfied by the correlation functions of the Markovian dynamics unless the dissipative part of the generator is zero. This result was shown by Talkner in his thesis [118], but in a formalism which was quite different from that used here. Therefore a proof of this result is included here.

The autocorrelation function for the unitary evolution of the closed system $S + R$ is conventionally defined as follows

$$r(X,Y;t) = \rho(U(t)^+XU(t)Y) = \rho(XU(t)YU(t)^+), \qquad (A.16)$$

where ρ is assumed to be invariant. Making the replacements

$$X \rightarrow X \otimes I, \qquad Y \rightarrow Y \otimes I, \qquad \rho \rightarrow \rho \otimes \rho_R$$

and defining the reduced dynamics for ρ through (A.1), we find that the autocorrelation function for S should be written in the following way

$$r(X,Y;t) = \rho(T(t)[X]Y), \quad \text{for } t > 0 \qquad \qquad \text{(A.17)}$$

$$= \rho(XT(-t)[Y]), \quad \text{for } t < 0,$$

where it is again assumed that the state ρ is invariant under the reduced dynamics

$$T(t)^*[\rho] \equiv \rho \cdot T(t) = \rho \quad , \text{ all } t > 0.$$

The function (A.16) has the positive semi-definite property

$$\sum_{k,1} r(X_k^+, X_1; t_k - t_1) \geq 0 \qquad \qquad \text{(A.18)}$$

for all $\{X_k \in B(H), t_k \in R\}_1^n$, all n. Note that

$$r(X,Y;t)^* = r(Y^+, X^+; -t).$$

If the reduced dynamics given by $T(t)$ has the CP and semigroup properties, then the property (A.18) holds also for (A.17). The proof is rather straightforward. Order the t_k: $t_1 \geq t_2 \geq \ldots \geq t_n$. As a matrix in $B(H_S) \otimes M_m$

$$X_{k,m}^+ \, X_{1,m} \geq 0$$

where $X_{1,m} = T(t_1 - t_m)[X_1]$, $1 \leq k,1 \leq m \leq n$. From the CP property of T follows a generalized Schwarz inequality: For $1 \leq k,1 \leq n-1$

$$T(t_{n-1} - t_n)[X_{k,n-1}^+ X_{1,n-1}] \geq X_{k,n}^+ X_{1,n}.$$

This is an immediate consequence of (A.3) and $T[I] = I$. Consequently the following operator-valued matrix is non-negative:

$$\left[\begin{array}{cc} T(t_{n-1} - t_n)[X_{k,n-1}{}^+X_{1,n-1}] & X_{n,n}{}^+X_{1,n} \\ X_{k,n}{}^+X_{n,n} & X_{n,n}{}^+X_{n,n} \end{array} \right] \geq 0.$$

Iteration of this inequality, using the semigroup property, and taking the expectation in the T-invariant state gives the desired relation (A.18). From this follows directly that the correlation function has a Fourier transform expression

$$r(X,Y;t) = \int de(X,Y;\omega) \exp(i\omega t)$$

where the spectral function has the positivity property, as a matrix function,

$$de(X_k^+, X_1; \omega) \geq 0, \quad \text{all } \omega,$$

$$de(X,Y;\omega)^* = de(Y^+, X^+; \omega).$$

The KMS condition can be written in the following way [87]: The correlation function can be continued to a function $r(X,Y;z)$ holomorphic in the strip $-\beta\hbar < \text{Im } z < 0$, and the boundary values of which satisfy

$$r(X,Y;t - i\beta\hbar) = r(Y,X; - t) \tag{A.19}$$

For the spectral function the equivalent condition reads

$$\exp(\beta\hbar\omega) \, de(Y,X; - \omega) = de(X,Y;\omega) \tag{A.20}$$

The KMS condition can not be satisfied when $T(t) = \exp(tL)$, where L is of the form (A.5) and ρ is faithful (and T-invariant), unless the dissipative part of L is zero. The proof is simple if L is assumed to be bounded. Then there is a convergent series expansion

$$T(z)[X] = \sum_n \frac{z^n}{n!} L^n[X],$$

which defines the RHS as an operator-valued holomorphic function in $|z| < \delta$ for $\delta = \|L\|^{-1}$. For bounded X,Y this means that

$$r_1(X,Y;z) \equiv \sum_n \frac{z^n}{n!} \rho(L^n[X]Y)$$

is a holomorphic function in $|z| < \delta$. By the KMS condition it can be continued to $- \beta\hbar < \text{Im } z < 0$. The boundary values of the analytic continuation for $t < 0$ must coincide with the boundary values of r there: $r_1(X,Y;t) = r(X,Y;t)$ for $t < 0$. Derivation at $t = 0$ gives

$$\rho(L[X]Y) = - \rho(X L[Y]). \tag{A.21}$$

Together with the invariance of ρ this means that

$$\rho(D(X^+,X)) = 0, \qquad \text{all } X \in B(H_S),$$

where D is the positive definite function of (A.6). As ρ is taken to be faithful, we obtain that $D(X^+,X) = 0$, which means that L is purely Hamiltonian. In the WCL one can alternatively use the detailed balance condition of A.1 to find for the invariant Gibbs state ρ that

$$\rho(L_d[X]Y) = \rho(X L_d[Y]),$$

which with (A.21) means that $L_d = 0$.

In the case where L is unbounded the correlation function is not necessarily differentiable at $t = 0$. It is then convenient to consider the resolvent of L , $R(\lambda) = (\lambda - L)^{-1}$, Re $\lambda > 0$, which has the familiar properties

$$R(\lambda) - R(\mu) = (\mu - \lambda)R(\mu)R(\lambda),$$

$$L \cdot R(\lambda) = \lambda R(\lambda) - I,$$

and which is related to the spectral function through

$$\frac{d}{d\omega} e(X,Y;\omega) = \lim_{\varepsilon \to 0} \frac{1}{2\pi} \rho(R(\varepsilon+i\omega)[X]Y + X R(\varepsilon-i\omega)[Y])$$

If we put $X_z = R(z)[X]$, it is easily found that for Re $z > 0$

$$D(X_z^+, X_z) = L[X_z^+ X_z] - (z + z^*)X_z^+ X_z + X^+ X_z + X_z^+ X,$$

$$\rho(X_z^+ X + X^+ X_z) = \rho(D(X_z^+, X_z)) + (z + z^*)\rho(X_z^+ X_z),$$

and consequently

$$\frac{d}{d\omega} e(X^+, X; \omega) =$$

$$= \frac{1}{2\pi} \lim_{\varepsilon \to 0} [\rho(D(X_z^+, X_z)) + (z + z^*)\rho(X_z^+ X_z)]_{z = \varepsilon - i\omega}$$

Due to (A.6) both terms in the RHS are non-negative. In order to have a finite spectral resolution the RHS must go to zero as $|\omega| \to \infty$. If X is such that this is true and $\lambda > 0$, then

$$\lim_{|\omega| \to \infty} \lim_{\varepsilon \to 0} [\rho(D(X_\lambda^+ - X_z^+, X_\lambda - X_z))]_{z = \varepsilon - i\omega}$$

$$= \rho(D(X_\lambda^+, X_\lambda))$$

$$\lim_{|\omega| \to \infty} \lim_{\varepsilon \to 0} [(z + z^*)\rho(R(z)[X_\lambda]^+ R(z)[X_\lambda])]_{z = \varepsilon - i\omega}$$

$$= 0$$

and we obtain the following asymptotic estimate for the spectral density as $|\omega| \to \infty$

$$\frac{d}{d\omega} e(X_\lambda^+, X_\lambda; \omega) \simeq \frac{1}{2\pi} (\omega^2 + \lambda^2)^{-1} \rho(D(X_\lambda^+, X_\lambda)).$$

The Cauchy measure contribution to the spectral function clearly contradicts the condition (A.20), which says that there is an exponential decrease in at least one direction.

Note that in the limit $\beta = 0$ (or $\hbar = 0$) there is no restriction on the correlation function.

A.5 Quantum thermal fluctuations

A fact which is closely related to the property described in A.4
is the lack of a "quantum white noise" [64-66]. The KMS condition
implies that the equilibrium thermal fluctuations can not have a
δ-function autocorrelation except in the limit $\beta = 0$, as the spectral function satisfying (A.20) can not be constant. Note that
there are other possible ideas of what one should mean by "white
noise" for quantum systems, as for example that described by
Maassen [119].

On the other hand, it is well known that it is only in the
limit where the reservoirs have infinitely short relaxation time
(the singular reservoir limit) that a semigroup evolution for the
open system can be derived without the time rescaling of the WCL
[64,65]. In fact, in this highly idealized limit the evolution
will be Markovian in the sense of (A.10), as the argument of [64]
can be repeated more or less verbatim also for a time-dependent
Hamiltonian (see also [116]). For finite-temperature reservoirs
the reduced evolution of the open system (without rescaling of the
time) must have a non-Markovian character, and no simple stochastic
reduced dynamics seems to be possible without contradicting the
thermodynamic consistency.

One aspect of the KMS property of the correlation functions,
which seems not to have been noticed before, is that they lack a
true irreversibility or unpredictability property. Take, for some
choice of X, $r(t) = r(X^+, X; t)$ as the autocorrelation function of
a complex-valued second order stationary stochastic process (see
e.g. Gihman and Skorohod [120] for a background). In the prediction
theory for such processes there is a decomposition of the process
into a singular part (which can be predicted uniquely from the
observation of the system in the past) and a regular part (which
is a filtered white noise and can not be predicted). The condition

148

that the process shall not be singular is that

$$J \equiv \int_{-\infty}^{\infty} d\omega \ (1 + \omega^2)^{-1} \ln f(\omega) > - \infty$$

where $f(\omega)d\omega$ is the absolutely continuous component of the spectral measure $e(\omega)$. From the KMS condition follows that

$$\ln f(- \omega) = - \beta \hbar \omega + \ln f(\omega).$$

Consequently the integral J can be written in the form

$$J = 2\int_{0}^{\infty} d\omega \ (1 + \omega^2)^{-1} \ln f(\omega) - \beta\hbar\int_{0}^{\infty} d\omega \ \omega(1 + \omega^2)^{-1}$$

As the first term of the RHS is $< \infty$, it follows that $J = - \infty$. Thus the classical stochastic process with autocorrelation $r(t)$ is singular and completely predictable. This means that in the limit $t \to \infty$, no additional information is obtained by observing the outcome. It is only in the limit $\beta = 0$ or $\hbar = 0$ that a truly nondeterministic process is obtained. The physical interpretation of this property is not evident, as the quantum nature of the observations on the system has not been taken into account, but it seems to be in line with the lack of mixing properties of Hamiltonian quantum dynamics mentioned in Chapter 8, though in this case it is not dependent on a discrete spectrum.

APPENDIX B. SENSITIVITY OF HYPERBOLIC MOTION

The object of this section is to show, by a simple but plausibly
typical example, that an exponentially unstable dynamical system
has a characteristic sensitivity to an external perturbation in
the form of an additive diffusion term. This property is expressed
in terms of the increase of the I-entropy of the state rather than
in terms of an exclusively classical concept like the trajectories
of the system. It is thus formulated in a way which may possibly
be generalized to quantum systems, at least in an approximate
fashion (see Chapter 8).

Consider the following diffusion process in R^n (compare [79])

$$dx(t) = \underline{\lambda} \cdot x(t)dt + \varepsilon \, dw(t) \qquad\qquad (B.1)$$

where $\underline{\lambda}$ is a constant real symmetric traceless $n \times n$ matrix, and w
represents the homogenous Wiener process on R^n. The deterministic
system corresponding to $\varepsilon = 0$, with solution

$$x(t) = \exp(t\underline{\lambda}) \cdot x(0),$$

where $\mathrm{Det}[\exp(t\underline{\lambda})] = 1$, has the sensitive dependence on initial
conditions of Anosov systems [73], provided that the eigenvalues
of $\underline{\lambda}$ are all non-zero. However, as there is no finite invariant
measure on the non-compact space R^n, this is not an abstract dynamical system according to conventional usage, and it can not describe the approach to equilibrium. Of course, similar models can
be constructed on compact spaces at some cost in more difficult
calculations. The lack of an equilibrium measure is not a serious
defect here, as the interest in the present calculation is in
states far from equilibrium.

(B.1) gives the following equation of evolution for the probability density on R^n (Δ = the Laplacian, ∇ = the gradient)

$$\frac{\partial f}{\partial t}(x,t) = \nabla f(x,t) \cdot \underline{\lambda} \cdot x + \frac{1}{2} \varepsilon^2 \Delta f(x,t). \qquad (B.2)$$

The solution can be written in the form

$$f(x,t) = \int_{R^n} dy\ G(\exp(t\underline{\lambda}) \cdot x - y, t) f(y,0)$$

$$G(x,t) = [\text{Det}(2\pi\varepsilon^2 \underline{r}(t))]^{-\frac{1}{2}} \exp[-\frac{1}{2} x \cdot (\varepsilon^2 \underline{r}(t))^{-1} \cdot x]$$

$$\underline{r}(t) = \int_0^t ds\ \exp(2s\underline{\lambda}).$$

As the evolution clearly maps the class of Gaussian densities into
itself, it is convenient to restrict the calculations to this class.
If the initial state is taken to be a Gaussian distribution with
covariance matrix $\underline{\sigma}$

$$f(x,0) = [\text{Det}(2\pi\underline{\sigma})]^{-\frac{1}{2}} \exp(-\frac{1}{2} x \cdot \underline{\sigma}^{-1} \cdot x),$$

where $\underline{\sigma}^{-1}$ is assumed to be non-singular, then

$$f(x,t) = [\text{Det}(2\pi\underline{\sigma}(t))]^{-\frac{1}{2}} \exp(-\frac{1}{2} x \cdot \underline{\sigma}(t)^{-1} \cdot x) \qquad (B.3)$$

$$\underline{\sigma}(t) = \exp(-t\underline{\lambda}) \cdot [\underline{\sigma} + \varepsilon^2 \underline{r}(t)] \cdot \exp(-t\underline{\lambda}) \qquad (B.4)$$

We use the notations $f_0, \underline{\sigma}_0$ to denote the solutions of (B.3) and
(B.4), respectively, when $\varepsilon = 0$.

The I-entropy of a continuous probability density is generally
not positive, but the relative entropy of two such densities is.
The relative entropy of two Gaussian distributions g_1, g_2 of the
same mean and with covariance matrices $\underline{\sigma}_1$, $\underline{\sigma}_2$, respectively, is
given by [93]

$$S_I(g_1|g_2) = \frac{1}{2} \text{Tr}[\ln(\underline{\sigma}_2 \cdot \underline{\sigma}_1^{-1}) + \underline{\sigma}_1 \cdot \underline{\sigma}_2^{-1} - 1]$$

A formal rearrangement gives from the definition

$$S_I(g_1|g_2) = S_I(g_2) - S_I(g_1) - \int dx \, (g_1 - g_2) \ln g_2 .$$

We can then define the difference in the I-entropies of the two distributions to be

$$\Delta S \equiv S_I(g_2) - S_I(g_1) = \frac{1}{2} \text{Tr} [\ln(\underline{\sigma}_2 \cdot \underline{\sigma}_1^{-1})] .$$

In order to obtain the increase of the I-entropy due to the diffusion term in (B.2), compared with the unperturbed evolution with $\varepsilon = 0$, replace g_2 by f and g_1 by f_0. Then we obtain

$$\Delta S(t) = \frac{1}{2} \text{Tr} [\ln(1 + \varepsilon^2 \underline{\sigma}^{-\frac{1}{2}} \cdot \underline{r}(t) \cdot \underline{\sigma}^{-\frac{1}{2}})] . \tag{B.5}$$

The asymptotic behaviour of this quantity when $t \to \infty$ follows directly from the expression for $\underline{r}(t)$:

$$\lim_{t \to \infty} t^{-1} \Delta S(t) = \sum_+ \lambda_j ,$$

where the sum is over all positive eigenvalues of $\underline{\lambda}$. There is thus an asymptotically constant rate of I-entropy production which is independent of ε. This unlimited growth of the I-entropy is due to the lack of an equilibrium distribution. Note the the assumption of a Gaussian initial state is not essential as all initial distributions will give the same asymptotic behaviour.

A closely related sensitivity property of more general dynamical systems on compact Riemannian manifolds was discussed by Kifer [121]. He compared the deterministic trajectories and the random paths obtained when a small diffusion term is added. Under some hyperbolicity condition the probability of the diffusion path remaining within a small δ of the unperturbed trajectory during $(0,t)$ is exponentially small as $t \to \infty$. The exponent is the sum of the positive Liapounov exponents (when these are defined) and it is again independent of ε. It is also interesting to recall Pesin's formula relating the KS entropy and the Liapounov exponents in this context [74]. We see that the hyperbolic systems have a sensitivity to perturbations which is expressed in terms of intrinsic properties (Liapounov exponents or KS entropy) and which is independent of the amplitude of the perturbations (in an asymptotic sense).

The limit treated above is that where the perturbation dominates the behaviour of the system. The opposite limit is where the diffusion can be neglected and a closed system description is permitted. The condition that the system is not significantly affected by the noise during the interval $(0,t)$ can be written in terms of the I-entropy: For some $\delta > 0$

$$\Delta S(t) \leq \delta .$$

From (B.5) follows that it is sufficient to demand that

$$\varepsilon^2 \text{Tr}[\underline{\sigma}^{-1} \cdot \underline{r}(t)] \leq 2\delta . \tag{B.6}$$

In the limit $\delta \to 0$ this is also a necessary condition. Introduce the spectral resolution of $\underline{\lambda}$:

$$\underline{\lambda} = \sum_j \lambda_j \underline{P}_j .$$

Then we find that

$$\text{Tr}[\underline{\sigma}^{-1} \cdot \underline{r}(t)] = \text{Tr}[\underline{\sigma}^{-1}] \sum_j p_j (2\lambda_j)^{-1} [\exp(2\lambda_j t) - 1]$$

$$\text{Tr}[\underline{\sigma}^{-1}] p_j \equiv \text{Tr}[\underline{\sigma}^{-1} \cdot \underline{P}_j] .$$

Jensen's inequality, applied to the concave function $\ln x$, gives

$$\ln(\text{Tr}[\underline{\sigma}^{-1} \cdot \underline{r}(t)]) \geq 2ht + 2k(t) + \ln(\text{Tr}[\underline{\sigma}^{-1}]),$$

$$h \equiv \sum_+ p_j \lambda_j ,$$

$$k(t) \equiv \frac{1}{2} \sum_j p_j \ln[(2|\lambda_j|)^{-1} (1 - \exp(-2t|\lambda_j|))]$$

h is a weighted average of the positive eigenvalues of the type occuring in Pesin's formula. $k(t)$ is a non-decreasing function of t and it satisfies

$$\lim_{t \to 0} (\ln t)^{-1} k(t) = \frac{1}{2} \, ,$$

$$\lim_{t \to \infty} k(t) = - \frac{1}{2} \sum_j p_j \ln(2|\lambda_j|) < \infty \, .$$

If (B.6) is satisfied, then the following weaker bound must hold for ε

$$\varepsilon \leq C \exp(-ht - k(t)) \tag{B.7}$$

$$C \equiv [2\delta / \text{Tr}(\underline{\sigma}^{-1})]^{\frac{1}{2}} .$$

If the dependence of h, k and C on the initial state is indicated, this inequality is (8.1). It is not evident that a similar bound can be written down for more general (non-Gaussian) initial states, and no attempt will be made here to treat the general case. However, it seems likely that the only special property which singles out the Gaussian distributions is their fast decrease at infinity. Initial states with a slow decrease at infinity may be less sensitive to the diffusion term, but then the non-compactness of the manifold is an artifice of the model. Hence, in order to have a more general treatment, a less artificial model is called for, i.e. one with a genuine equilibrium distribution.

Another artificial property of the model is that $\underline{\lambda}$ is a constant. A more realistic assumption is that $\underline{\lambda}$ is a function of x. The exponential instability in different points can then cancel in the integrated motion to produce a regular, stable behaviour [80-83]. This effect can be mimicked in the simple model (B.2) by having a time-dependent $\underline{\lambda}$. From (B.4) follows that

$$\frac{d}{dt} \underline{\sigma}(t) = - \underline{\lambda}(t) \cdot \underline{\sigma}(t) - \underline{\sigma}(t) \cdot \underline{\lambda}(t) + \varepsilon^2 \underline{I} \tag{B.8}$$

If $\underline{u}(t)$ is the unique solution of

$$\frac{d}{dt} \, \underline{u}(t) = \underline{\lambda}(t) \cdot \underline{u}(t), \qquad \underline{u}(0) = \underline{I} \ ,$$

and if we put

$$\underline{r}(t) = \int_0^t ds \, \underline{u}(s)^+ \cdot \underline{u}(s) \ ,$$

then the solution of (B.8) is given by

$$\underline{\lambda}(t) = (\underline{u}(t)^+)^{-1} \cdot [\underline{\sigma}(0) + \varepsilon^2 \underline{r}(t)] \cdot \underline{u}(t)^{-1}$$

If the unperturbed evolution

$$\frac{d}{dt} \, x(t) = \underline{\lambda}(t) \cdot x(t)$$

has solutions of a common period τ, then $\underline{u}(t)$ has period τ and $\underline{r}(t)$ is of the asymptotic form (valid for $t \gg \tau$)

$$\lim_{t \to \infty} t^{-1} \underline{r}(t) = \underline{r}_0$$

The same asymptotic form holds if $\underline{u}(t)$ is quasiperiodic. Inserted into (B.5) this behaviour implies that

$$\lim_{t \to \infty} (\ln t)^{-1} \Delta S(t) = \frac{1}{2} \ , \tag{B.9}$$

rather than the asymptotic linear increase of the I-entropy valid for the exponentially unstable case treated above. The bound on ε given by (B.6) now reads (for $t \gg \tau$)

$$\varepsilon^2 \leq 2\delta (t \, Tr[\underline{\sigma}^{-1} \cdot \underline{r}_0])^{-1} \tag{B.10}$$

instead of (B.7). This bound is of the form (8.2).

REFERENCES

1. O. Penrose: Rep. Prog. Phys. 42, 1937 (1979)
2. P.C.W. Davies: *The physics of time asymmetry*.
    Surrey University Press 1974
3. D. Wolf: *Spin-temperature and nuclear-spin relaxation in matter*. Oxford University Press 1979
4. R. Lenk: Prog. Nucl. Magn. Resonance Spectrosc. 13, 271 (1979)
5. W.K. Rhim, A. Pines, J.S. Waugh: Phys. Rev. B3, 684 (1971)
6. W.K. Rhim, D.D. Elleman, R.W. Vaughan: J. Chem. Phys. 59, 3740 (1973)
7. E.T. Jaynes: Am. J. Phys. 33, 391 (1965)
8. M. Reed, B. Simon: *Methods of modern mathematical physics*, Vol. 2 (§ X.12) New York: Academic Press 1975
9. K. Yosida: *Functional Analysis*. Berlin: Springer 1965
10. H. Tanabe: *Equations of evolution*. London: Pitman 1979
11. B. Mielnik: J. Math. Phys. 21, 44 (1980)
12. J. Waniewski: Commun. Math. Phys. 76, 27 (1980)
13. G. Lindblad: Commun. Math. Phys. 48, 119 (1976)
14. V. Gorini, A. Kossakowski, E.C.G. Sudarshan: J. Math. Phys. 17, 821 (1976)
15. V. Gorini, A. Frigerio, M. Verri, A. Kossakowski, E.C.G. Sudarshan: Rep. Math. Phys. 13, 149 (1978)
16. A. Wehrl: Rev. Mod. Phys. 50, 221 (1978)
17. W. Thirring: *Lehrbuch der Mathematischen Physik*, Vol. 4. Wien: Springer 1980
18. W. Ochs: Rep. Math. Phys. 8, 109 (1975)
19. G. Lindblad: Commun. Math. Phys. 33, 305 (1973)
20. G. Lindblad: Commun. Math. Phys. 39, 111 (1974)
21. G. Lindblad: Commun. Math. Phys. 40, 147 (1975)

22. E.H. Lieb: Advances in Math. 11, 267 (1973)

23. E.H. Lieb, M.B. Ruskai: J. Math. Phys. 14, 1938 (1973)

24. H. Spohn, J.L. Lebowitz: Adv. Chem. Phys. 38, 109 (1978)

25. G. Lindblad: A general H-theorem for quantum stochastic processes. Report TRITA-TFY-79-21, Stockholm 1979

26. W. Pusz, S.L. Woronowicz: Commun. Math. Phys. 58, 273 (1978)

27. I. Procaccia, R.D. Levine: J. Chem. Phys. 65, 3357 (1976)

28. G.L. Sewell: Phys. Reports 57(5), 307 (1980)

29. S. Goldstein, O. Penrose: J. Stat. Phys. 24, 325 (1981)

30. H. Primas: Helv. Phys. Acta 34, 36 (1961)

31. A. Kossakowski: Rep. Math. Phys. 3, 247 (1972)

32. R.C. Tolman: The principles of statistical mechanics (Ch 12, 13). Oxford University Press 1938

33. J.M. Blatt: Prog. Theor. Phys. 22, 745 (1959)

34. N.S. Krylov: Works on the foundations of statistical mechanics. Princeton University Press 1979

35. O.E. Lanford, J.L. Lebowitz: Time evolution and ergodic properties of harmonic systems. In: J. Moser (ed.), Lecture Notes in Physics 38. Berlin: Springer 1975

36. J.L. van Hemmen: Phys. Reports 65(2), 43 (1980)

37. E.B. Davies: Quantum theory of open systems. London: Academic Press 1976

38. E.B. Davies: Commun. Math. Phys. 39, 91 (1974)

39. E.B. Davies: Math. Ann. 219, 147 (1976)

40. H. Spohn: Rev. Mod. Phys. 53, 569 (1980)

41. E.B. Davies, H. Spohn: J. Stat. Phys. 19, 511 (1978)

42. A. Frigerio: Lett. Math. Phys. 2, 79 (1977)

43. A. Frigerio, V. Gorini, M. Verri: Stability, detailed balance and KMS condition for quantum systems. In: G. Dell'Antonio, S. Doplicher, G. Jona-Lasinio (eds.), Lecture Notes in Physics 80. Berlin: Springer 1978

44. A. Kossakowski, A. Frigerio, V. Gorini, M. Verri: Commun. Math. Phys. 57, 97 (1977)

45. R. Haag, D. Kastler, E. Trych-Pohlmeyer: Commun. Math. Phys. 38, 173 (1974)

46. G. Benettin, G. Lo Vecchio, A. Tenenbaum: Phys. Rev. A22, 1709 (1980)

47. H. Narnhofer, W. Thirring: Phys. Rev. A26, 3646 (1982)

48. T. Erber, B. Schweitzer, A. Sklar: Commun. Math. Phys. 29, 311 (1973)

49. Ya. G. Sinai: *Introduction to ergodic theory* (Ch 9,10). Princeton University Press 1977

50. J. Meixner: *Linear passive systems*. In: J. Meixner (ed.) *Statistical mechanics of equilibrium and non-equilibrium*. Amsterdam: North Holland 1965

51. J. Mehra, E.C.G. Sudarshan: Nuovo Cimento 11B, 215 (1972)

52. R.S. Ingarden, A. Kossakowski: Ann. Phys. (NY) 89, 451 (1975)

53. H. Spohn: J. Math. Phys. 19, 1227 (1978)

54. R. Alicki: J. Phys. A12, L103 (1979)

55. R. Alicki: J. Stat. Phys. 20, 671 (1979)

56. G. Lindblad: Commun. Math. Phys. 65, 281

57. G. Lindblad: *Response of Markovian and non-Markovian quantum stochastic systems to time-dependent forces*. Report TRITA-TFY-79-9, Stockholm 1979

58. C.P. Slichter: *Principles of magnetic resonance*. Berlin: Springer 1978

59. O. Penrose: *Foundations of statistical mechanics* (§§ I.3, V.1). Oxford: Pergamon Press 1970

60. M. Reed, B. Simon: *Methods of modern mathematical physics* Vol. 1 (§ VIII.8). New York: Academic Press 1972

61. L.E. Reichl: *A modern course in statistical physics*. E. Arnold 1980

62. U. Haeberlen: *High resolution NMR in solids*. Adv. Magn. Res. Supplement 1 (1976)

63. G. Lindblad: *Quantum stochastic processes on matrix algebras*. Report TRITA-TFY-78-10, Stockholm 1978

64. V. Gorini, A. Kossakowski: J. Math. Phys. 17, 1298 (1976)
65. A. Frigerio, V. Gorini: J. Math. Phys. 17, 2123 (1976)
66. R. Kubo: J. Phys. Soc. Japan 26, Supplement, 1 (1969)
67. E. Borel: *Le Hasard*. Paris: Felix Alcan 1914
68. E. Borel: *Traité du calcul des probabilités et de ses appli-cations*. Tome 2, Fasc. 3: *Mechanique statistique classique*. Paris: Gauthier-Villars 1925
69. M.V. Berry: *Regular and irregular motion*. In: S. Jorma (ed.) Am. Inst. Phys. Conf. Proc. 46 (1978)
70. Ya.G. Sinai: Acta Phys. Austriaca Suppl. 10, 575 (1973)
71. D. Ruelle: Ann. N.Y. Acad. Sci. 316, 408 (1979)
72. R. Shaw: Z. Naturforsch. 36a, 80 (1981)
73. V.I. Arnold, A. Avez: *Ergodic problems of classical mechanics*. New York: W.A. Benjamin 1968
74. Ya.B. Pesin: Russ. Math. Surveys 32:4, 55 (1979)
75. G.G. Emch: Commun. Math. Phys. 49, 191 (1976)
76. A. Connes, E. Størmer: Acta Math: 134, 289 (1975)
77. A. Connes, E. Størmer: *A connection between the classical and the quantum mechanical entropies*. Preprint, Dept. of Mathe-matics, Oslo University 1980
78. Ya.G. Sinai: Russ. Math. Surveys 27:4, 21 (1972)
79. Yu.I. Kifer: Theory Prob. Appl. 19, 487 (1974)
80. M. Tabor: Adv. Chem. Phys. 46, 73 (1981)
81. G. Benettin, R. Brambilla, L. Galgani: Physica 87A, 381 (1977)
82. G. Casati, B.V. Chirikov, J. Ford: Phys. Lett. 77A, 91 (1980)
83. I. Hamilton, P. Brumer: Phys. Rev. A23, 1941 (1981)
84. D.W. Noid, M.L. Koszykowski, R.A. Markus: Ann. Rev. Phys. Chem. 32, 267 (1981)
85. G.M. Zaslavsky: Phys. Reports 80(3), 157 (1981)
86. I.C. Percival: J. Phys. B6, L229 (1973)
87. D. Ruelle: *Statistical mechanics* (Ch 7). New York: W.A. Benjamin 1969

88. A. Martin-Löf: *Statistical mechanics and the foundations of thermodynamics*. Lecture Notes in Physics $\underline{101}$. Berlin: Springer 1979

89. J.P. Eckmann: Rev. Mod. Phys. $\underline{54}$, 643 (1981)

90. P.C. Hemmer, L.C. Maximon, H. Wergeland: Phys. Rev. $\underline{111}$, 689 (1958)

91. P. Mazur, E. Montroll: J. Math. Phys. $\underline{1}$, 70 (1960)

92. A. Peres: Phys. Rev. Lett. $\underline{49}$, 1118 (1982)

93. G. Lindblad: J. Stat. Phys. $\underline{11}$, 231 (1974)

94. R. Kosloff: Adv. Chem. Phys. $\underline{46}$, 153 (1981)

95. L. Szilard: Z. Phys. $\underline{53}$, 840 (1929)

96. L. Brillouin: *Science and information theory*. New York: Academic Press 1956

97. R.P. Poplavskii: Sov. Phys. Usp. $\underline{22}$, 371 (1979)

98. J.M. Jauch, J.G. Baron: Helv. Phys. Acta $\underline{45}$, 220 (1972)

99. K. Popper: *Intellectual autobiography* (§ 36). In: P.A. Schilpp (ed.) *The philosophy of Karl Popper*. The library of living philosophers Vol XIV. La Salle, Ill: Open Court 1974

100. C.W.F. McClare: J. Theor. Biol. $\underline{30}$, 1 (1971)

101. M. Planck: *Thermodynamics* (§ 112). New York: Dover 1945

102. I. Prigogine, C. George, F. Henin, L. Rosenfeld: Chem. Scr. $\underline{4}$, 5 (1973)

103. B. Misra: Proc. Natl. Acad. Sci. USA $\underline{75}$, 1627 (1978)

104. B. Misra, I. Prigogine, M. Courbage: Physica $\underline{98A}$, 1 (1979)

105. B. Misra, I. Prigogine: Suppl. Progr. Theor. Phys. $\underline{69}$, 101 (1980)

106. S. Goldstein, B. Misra, M. Courbage: J. Stat. Phys. $\underline{25}$, 111 (1981)

107. J.E. Mayer, M.G. Mayer: *Statistical mechanics*, 2nd Ed (Ch 6) New York: Wiley 1976

108. O.E. Lanford: Acta Phys. Austriaca Suppl. $\underline{10}$, 619 (1973)

109. S. Goldstein: Israel J. Math. $\underline{38}$, 241 (1981)

110. P.W. Bridgman: *The nature of thermodynamics* (p 168).
Cambridge: Harvard University Press 1943

111. H. Grad: *Levels of description in statistical mechanics and thermodynamics*. In: M. Bunge (ed.) *Delaware seminar on the foundations of physics*, Vol. 1. Berlin: Springer 1967

112. E.T. Jaynes: *Were do we stand on maximum entropy?* In:
R.D. Levine, M. Tribus (eds.) *The maximum entropy formalism*.
Cambridge: MIT Press 1979

113. B.D. Coleman, D.R. Owen: Arch. Rational Mech. Anal. $\underline{54}$, 1 (1974)

114. W.A. Day: Arch. Rational Mech. Anal. $\underline{31}$, 1 (1968)

115. M. Rosenblatt: *Markov processes* (Ch 3). Berlin: Springer 1971

116. R. Dümcke: J. Math. Phys. $\underline{24}$, 311 (1983)

117. A.G. Redfield: Phys. Rev. $\underline{98}$, 1787 (1955)

118. P. Talkner: *Untersuchungen irreversibler Prozesse in quanten-mechanischen Systemen*. Thesis, Stuttgart 1979

119. H. Maassen: *On a class of quantum Langevin equations and the question of approach to equilibrium*. Thesis, Groningen 1982

120. I.I. Gikhman, A.V. Skorohod: *Introduction to the theory of random processes* (Ch 5). Philadelphia: W.B. Saunders 1965

121. Yu. Kifer: *A probabilistic version of Bowen - Ruelle's lemma*.
In: Z. Nitecki, C. Robinson (eds.) Lecture Notes in Mathematics $\underline{819}$. Berlin: Springer 1981

NOTATION INDEX

SUBJECT INDEX